The Maths Handbook

Everyday Maths Made Simple

Richard Elwes

Quercus

Contents

Introduction

'I was never any good at mathematics.'

I must have heard this sentence from a thousand different people.

I cannot dispute that it may be true: people do have different strengths and weaknesses, different interests and priorities, different opportunities and obstacles. But, all the same, an understanding of mathematics is not something anyone is born with, not even Pythagoras himself. Like all other skills, from portraiture to computer programming, from knitting to playing cricket, mathematics can only be developed through practice, that is to say through actually *doing it*.

Nor, in this age, is mathematics something anyone can afford to ignore. Few people stop to worry whether they are *good* at talking or *good* at shopping. Abilities may indeed vary, but generally talking and shopping are unavoidable parts of life. And so it is with mathematics. Rather than trying to hide from it, how about meeting it head on and becoming good at it?

Sounds intimidating? Don't panic! The good news is that just a handful of central ideas and techniques can carry you a very long way. So, I am pleased to present this book: a no-nonsense guide to the essentials of the subject, especially written for anyone who 'was never any good at mathematics'. Maybe not, but it's not too late!

Before we get underway, here's a final word on philosophy. Mathematical education is split between two rival camps. Traditionalists brandish rusty compasses and dusty books of log tables, while modernists drop fashionable buzzwords like 'chunking' and talk about the 'number line'. This book has no loyalty to either group. I have simply taken the concepts I consider most important, and illustrated them as clearly and straightforwardly as I can.

Many of the ideas are as ancient as the pyramids, though some have a more recent heritage. Sometimes a modern presentation can bring a fresh clarity to a tired subject; in other cases, the old tried and tested methods are the best.

Richard Elwes

The language of mathematics

- *Writing mathematics*
- *Understanding what the various mathematical symbols mean, and how to use them*
- *Using BEDMAS to help with calculations*

Let's begin with one of the commonest questions in any mathematics class:* 'Can't I just use a calculator?' *The answer is . . . of course you can! This book is not selling a puritanical brand of mathematics, where everything must be done laboriously by hand, and all help is turned down. You are welcome to use a calculator for arithmetic, just as you can use a word-processor for writing text. But handwriting is an essential skill, even in today's hi-tech world. You can use a dictionary or a spell-checker too. All the same, isn't it a good idea to have a reasonable grasp of basic spelling?

There may be times when you don't have a calculator or a computer to hand. You don't want to be completely lost without it! Nor do you want to have to consult it every time a few numbers need to be added together. After all, you don't get out your dictionary every time you want to write a simple phrase.

So, no, I don't want you to throw away your calculator. But I would like to change the way you think about it. See it as a labour saving device, something to speed up calculations, a provider of handy shortcuts.

The way I *don't* want you to see it is as a mysterious black box which performs near-magical feats that you alone could never hope to do. Some of the quizzes will show this icon (), which asks you to have a go without a calculator. This is just for practice, rather than being a point of principle!

Signs and symbols

Mathematics has its own physical toolbox, full of calculators, compasses and protractors. We shall meet these in later chapters. Mathematics also comes with an impressive lexicon of terms, from 'radii' to 'logarithms', which we shall also get to know and love in the pages ahead.

Perhaps the first barrier to mathematics, though, comes before these: it is the library of signs and symbols that are used. Most obviously, there are the symbols 0, 1, 2, 3, 4, 5, 6, 7, 8, 9. It is interesting that once we get to the number ten there is not a new symbol. Instead, the symbols for 0 and 1 are recycled and combined to produce the name '10'. Instead of having one symbol alone, we now have two symbols arranged in two *columns*. Which column the symbol is in carries just as much information as the symbol itself: the '1' in '13' does not only mean 'one', it means 'one ten'. This method of representing numbers in columns is at the heart of the *decimal system*: the modern way of representing numbers. It is so familiar that we might not realize what an ingenious and efficient system it is. Any number whatsoever can be written using only the ten symbols 0–9. It is easy to read too: you don't have to stop and wonder how much '41' is.

This way of writing numbers has major consequences for the things that we do with them. The best methods for addition, subtraction, multiplication and division are based around understanding how the columns affect each other. We will explore these in depth in the coming chapters.

There are many other symbols in mathematics besides numbers themselves. To start with, there are the four representing basic arithmetical procedures: $+, -, \times, \div$. In fact there are other symbols which mean the same things. In many situations, scientists prefer a dot, or even nothing at all, to indicate multiplication. So, in algebra, both ab and $a \cdot b$, mean the same as $a \times b$, as we shall see later. Similarly, division is just as commonly expressed by $\frac{a}{b}$ as by $a \div b$.

This use of letters is perhaps the greatest barrier to mathematics. How can you multiply and divide letters? (And why would you want to?) These are fair questions, which we shall save until later.

Writing mathematics

Here is another common question:

'What is the point of writing out mathematics in a longwinded fashion? Surely all that matters is the final answer?'

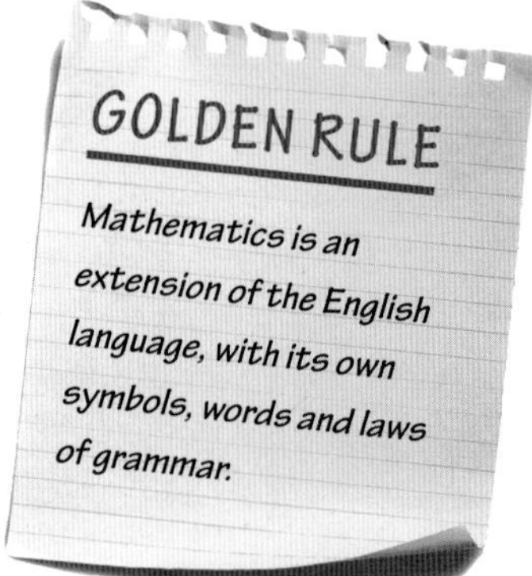

The answer is . . . no! Of course, the right answer is important. I might even agree that it is usually the most important thing. But it is certainly not the only important thing. Why not? Because you will have a much better chance of reliably arriving at the right answer if you are in command of the *reasoning* that leads you there. And the best way of ensuring *that* is to write out the intermediate steps, as clearly and accurately as possible.

Writing out mathematics has two purposes. Firstly it is to guide and illuminate your own thought-processes. You can only write things out clearly if you are thinking about them clearly, and it is this clarity of thought that is the ultimate aim. The second purpose is the same as for almost any other form of writing: it is a form of communication with another human being. I suggest that you work under the assumption that someone will be along shortly to read your mathematics (whether or not this is actually true). Will they be able to tell what you are doing? Or is it a jumble of symbols, comprehensible only to you?

Mathematics is an extension of the English language (or any other language, but we'll stick to English!), with some new symbols and words. But all the usual laws of English remain valid. In particular, when you write out mathematics, the aim should be prose that another person can read and understand. So try *not* to end up with symbols scattered randomly around the page. That's fine for rough working, while you are trying to figure out what it is you want to write down. But after you've figured it out, try to write everything clearly, in a way that communicates what you have understood to the reader, and helps them understand it too.

The importance of equality

The most important symbol in mathematics is '$=$'. Why? Because the number-one goal of mathematics is to discover the value of unknown quantities, or to establish that two superficially different objects are actually one and the same. So an *equation* is really a sentence, an assertion. An example is '$146 + 255 = 401$', which states that the value on the left-hand side of the '$=$' sign is the same as the value on the right.

It is amazing how often the '$=$' sign gets misused! If asked to calculate $13 + 12 + 8$, many people will write '$13 + 12 = 25 + 8 = 33$'. This may come from the use of calculators where the $\boxed{=}$ button can be interpreted to mean 'work out the answer'. It may be clear what the line of thought is, but taken at face value it is nonsense: $13 + 12$ is not equal to $25 + 8$! A correct way to write this would be '$13 + 12 + 8 = 25 + 8 = 33$'. Now, every pair of quantities that are asserted to be equal really are equal – a great improvement!

The '$=$' sign has some lesser-known cousins, which make less powerful assertions: '$<$' and '$>$'. For example, the statement '$A < B$' says that the quantity *A is less than B*. An example might be $3 + 9 < 13$. Flipping this around gives '$B > A$', which says that *B is greater than A*, for example, $13 > 3 + 9$. The statements '$A < B$' and '$B > A$' look different, but have exactly the same meanings (in the same way that '$A = B$' and '$B = A$' mean essentially the same thing).

Other symbols in the same family are '$\geq$' and '$\leq$', which stand for 'is greater than or equal to' and 'is less than or equal to' (otherwise known as 'is at least' and 'is at most').

In coming chapters, we will look at techniques for addition, subtraction, multiplication, division, and much else besides, which will allow us to judge whether or not these types of assertion are true.

Now we will have a look at one of the hidden laws of mathematical grammar.

A profusion of parentheses

HAVE A GO AT QUIZ 1.

One thing you may see in this book, which you may not be used to, is lots of brackets in among the numbers. Why is that? Rather than answering that question directly, I'll pose another. What is $3 \times 2 + 1$? At first sight, this seems easy enough. The trouble is that there are two ways to work it out:

a) $3 \times 2 + 1 = 6 + 1 = 7$

b) $3 \times 2 + 1 = 3 \times 3 = 9$

Only one of these can be right, but which is it?

To avoid this sort of confusion, it is a good idea to use brackets to mark out which calculations should be taken together. So the two above would be written like this:

a) $(3 \times 2) + 1$

b) $3 \times (2 + 1)$

Now both are unambiguous, and whichever one was intended can be written without any danger of misunderstanding. In each case, the first step is to work out the calculation inside the brackets.

The same thing applies with more advanced topics, such as negative numbers and powers. In the coming chapters we shall see expressions such as -4^2. But does this mean $-(4^2)$, that is to say -16, or does it mean $(-4)^2$, which as we shall see in the theory of negative numbers, is actually $+16$?

NOW HAVE A GO AT QUIZ 2.

BEDMAS

You might protest that I haven't answered the question at the start of the last section. Without writing in any brackets, what is $3 \times 2 + 1$?

There is a convention which has been adopted to resolve ambiguous situations like this. We can think of it as one of the grammatical laws of mathematics. It is called BEDMAS (or sometimes BIDMAS or BODMAS). It tells us the order in which the operations should be carried out:

Brackets **E**xponents **D**ivision **M**ultiplication **A**ddition **S**ubtraction

If you prefer, 'Exponents' can be replaced by 'Indices', giving BIDMAS (or with 'Orders', giving BODMAS). All of these options are words for *powers*, which we shall meet in a later chapter. (Unfortunately BPDMAS isn't quite as catchy.)

The point of this is that the order in which we calculate things follows the letters in 'BEDMAS'. In the case of $3 \times 2 + 1$, the two operations are multiplication and addition. Since M comes before A in BEDMAS, multiplication is done first, and we get $3 \times 2 + 1 = 6 + 1 = 7$ as the correct answer.

TIME FOR BEDMAS? HAVE A GO AT QUIZ 3AND 4

When we come to -4^2, the two operations are subtraction (negativity, to be pernickety) and exponentiation. Since E comes before S, the correct interpretation is $-(4^2) = -16$.

Calculators use BEDMAS automatically: if you type in $\boxed{3}\boxed{\times}\boxed{2}\boxed{+}\boxed{1}\boxed{=}$, you will get the answer 7 not 9.

Sum up *The way we think about life comes across in the way we talk and write about it. The same is true of mathematics. If you want your thought-processes to be clear and accurate, then start by focusing on the language you use!*

Quizzes

1 Translate these sentences into mathematical symbols, and decide whether the statement is true or false.

a When you add eleven to ten you get twenty-one.
b Multiplying two by itself gives the same as adding two to itself.
c When you subtract four from five you get the same as when you divide two by itself.
d Five divided by two is at least three.
e Five multiplied by four is less than three multiplied by seven.

2 Put brackets in these expressions in two different ways, and then work out the two answers. (For example from $3 \times 2 + 1$, we get $(3 \times 2) + 1 = 7$ and $3 \times (2 + 1) = 9$.)

a $1 + 2 + 3$ **b** $4 + 6 \div 2$ **c** $2 \times 3 \times 4$
d $20 - 6 \times 3$ **e** $2 \times 3 + 4 \times 5$

3 In each of the expressions in quiz 2, decide which is the correct interpretation according to BEDMAS. (If it doesn't matter, explain why.)

4 As well as BEDMAS, there is a convention that operations are read from left to right. So $8 \div 4 \div 2$ means $(8 \div 4) \div 2$ not $8 \div (4 \div 2)$. For which of addition, subtraction, multiplication, and division is this rule necessary?

Addition

- *Mastering simple sums*
- *Knowing how to 'carry' and borrow*
- *Remembering shortcuts for mental arithmetic*

Everyone knows what addition means: if you have 7 greyhounds and 5 chihuahuas, then your total number of dogs is 7 + 5. The difficulty is not in the meaning of the procedure, but in calculating the answer. The simplest method of all is to start at 7, and then add on 1 five times in succession. This might be done by counting up from 7 out loud: 8, 9, 10, 11, 12, keeping track by counting up to 5 on your fingers.

But counting up is much too slow! When large numbers are involved, such as 2789 + 1899, this technique would take several hours, and the likelihood of slipping up somewhere is close to certain. So how can this be speeded up? There are many different procedures which work well, depending on the context, and the quantity and types of numbers we are dealing with. We will have a look at several methods in this chapter.

The key thing is to be comfortable adding up the small numbers: those between 1 and 9. Once you can do this without worrying about it, then building up to larger and more complex sums becomes surprisingly easy.

GET UNDERWAY WITH QUIZ 1.

The aim here is not just to arrive at the right answers, but to be able to handle these types of calculation quickly and painlessly. If you feel you could do with more practice, then set yourself five questions at a time and work through them. Start as slowly as you like, and aim to build up speed with practice.

When numbers grow up

It is no surprise that addition becomes trickier when it involves numbers more than one digit long. So it is the length of the numbers that we have to learn to manage next.

Suppose we are faced with the calculation 20 + 40. This seems easy. But why? Because all that really needs to be done is to work out 2 + 4 and then stick a zero on the end. In the same way, even three-, four-, or five-digit numbers can be easy to handle: 3000 + 6000, for instance.

Things get slightly trickier when we have something like 200 + 900. Here, although the question involves only three-digit numbers, the answer steps up to four digits, just as $2 + 9$ steps up from one digit to two.

ZEROS AND VILLAINS? HAVE A GO AT QUIZ 2.

Numbers with a lot of zeros are the first kind of longer numbers to get used to.

Totalling columns

This chapter's golden rule tells us how to tackle longer numbers: arrange them in columns. The number '456', for example, needs three columns. It has a 6 in the *units column*, a 5 in the *tens column*, and a 4 in the *hundreds column*:

HUNDREDS	TENS	UNITS
4	5	6

Notice that we work along the columns from right to left, always beginning with the units. (The reason for this backwards approach will become clear later on.) Now suppose we want to add 456 to another number, say 123. The process is as follows. First write the two numbers out in columns, with one under the other. Make sure that the units in the top number are aligned with the units in the number below, and similarly for the tens and hundreds columns.

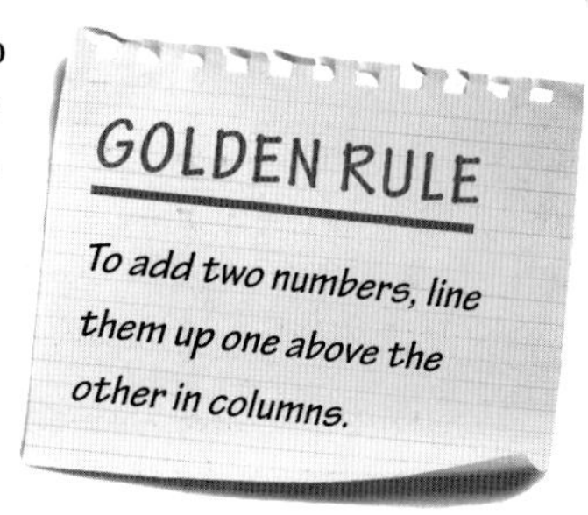

With that done, all that remains is to add up the numbers in each column:

	HUNDREDS	TENS	UNITS
	4	5	6
+	1	2	3
=	5	7	9

GOT THAT? THEN HAVE A GO AT QUIZ 3!

The art of carrying

Now we arrive at the moment where all the beautiful simplicity of the previous examples turns into something a bit more complex. At this stage the columns are no longer summed up individually, but start affecting each other through a mystifying mechanism known as *carrying*. I promise it isn't as bad as it sounds!

Let's start with an example: 44 + 28. What happens if we simply follow the procedure described in the last section?

	TENS	UNITS
	4	4
+	2	8
=	6	12

This is completely, 100%, correct! There is just one small worry: 'sixty-twelve' is not the name of any number in English. (Saying it might attract strange looks in the street.) So what is sixty-twelve in ordinary language? A little reflection should convince you that the answer is seventy-two. (In the French language, sixty-twelve, or *soixante douze,* is in fact the name for seventy-two.)

So, to complete the calculation, we need to rewrite the answer in the ordinary way, as 72. What exactly is going on in this final step? The answer is that the units column contains 12, which is too many. When we reduce 12 to 2, we are left with one extra ten to manage. It is this 1 (ten) which is 'carried' to the tens column.

Numbers are only ever carried leftwards: from the units column to the tens, or from the tens to the hundreds. (This is the reason we always work from right to left when adding numbers up.) Once we have grasped this essential idea, we can speed up the process by doing all the carrying as we go along.

So, let's take another example: 37 + 68. Here we begin by adding up the units column to get 15, which we can immediately write as 5 and carry the leftover ten as an extra 1 to be included in the tens column. (We write this as an extra 1 at the top of the column.) Then we add up the tens column (including the carried 1) which produces 10. So we write this as 0, and carry 1 to the hundreds column. Happily there is nothing else in the hundreds column, so this is the end.

	1	1	
		3	7
+		6	8
=	1	0	5

Now try this yourself in Quiz 4.

Lists of numbers

Whether it's counting calories, or adding up shopping bills, addition is probably our commonest use of numbers. But often the calculation needs more than two numbers to be added together. The good news is that the technique we learnt in the last section transfers immediately to longer lists of numbers. The rules are exactly the same as before: arrange the numbers vertically, and then add each column in turn, starting on the right, carrying when necessary. The only difference is that the number to be carried this time might be larger than 1.

For example, to calculate 36 + 27 + 18 we set it up as:

	3	6
	2	7
+	1	8
=		

This time the units column adds up to 21, so we write 1, and carry 2 to the tens column. Then we add up the tens column as before, to get 8.

IF THAT SEEMS MANAGEABLE, THEN TRY QUIZ 5.

In your head: splitting numbers up

The addition techniques we have looked at so far work very well (after a little practice). But they do have one downside: these are written techniques. Often what we want is a way to calculate in our head, without having to scuttle off to a quiet corner with a pen and paper. Carrying can be tricky to manage in your head. Luckily there are other ways to proceed.

If we want to add 24 to 51, one way to proceed is to split this up into two simpler sums: first add on 20, and then add on another 4. Each of these steps should be easy to do: 51 + 20 = 71 (because 5 + 2 = 7 in the tens column). Then 71 + 4 = 75. The only challenge is to keep a mental hold of the intermediate step (71 in this example).

Remember that you can choose which of the two numbers to split up. So we could have done the previous example as 24 + 50 + 1. You might find it better to split up the smaller of the two, but tastes vary.

TRY THIS YOURSELF IN QUIZ 6.

Rounding up and cutting down

Imagine that a restaurant bill comes to £45 for food, with another £29 for drink. By now we have seen a few techniques we could use to tackle the resulting sum: 45 + 29. But there is another possibility, which begins by noticing that 29 is 1 less than 30. So, to make life easier, we could round 29 up to 30. Then it is not hard to add 30 to 45 to get 75. To complete the calculation, we just need to cut it back down by 1 again, to arrive at 74.

This trick of rounding up and cutting down will also work when adding, say, 38 to 53. Instead of tackling the sum head-on, first round 38 up by 2, then add 40 to 53. To finish off, just cut that number back down by 2.

In some cases you might want to round up both numbers in the sum. For example, 59 + 28 can be rounded up to 60 + 30, and then cut down by a total of 3.

I think rounding up and cutting down is a good technique when the units column contains a 7, 8 or 9 and splitting numbers up is better when the units column contains a 1, 2 or 3. But it is up to you to decide which approaches suit you best! So why not try both techniques?

Sum up *Mathematics can teach us several techniques for addition and subtraction. But all of them are based on familiarity with the small numbers, 1 to 9.*

Quizzes

After you have worked through these, come up with your own examples if you want more practice. No calculators for this chapter!

1 In your head!

a 3 + 8
b 7 + 6
c 9 + 9
d 5 + 4 + 3
e 8 + 7 + 6

2 Numbers that grow longer

a 30 + 40
b 5000 + 2000
c 800 + 300
d 7000 + 4000
e 30,000 + 90,000

3 Write in columns and add.

a 56 + 22
b 48 + 51
c 195 + 503
d 354 + 431
e 1742 + 8033

4 Mastering carrying

a 14 + 27
b 36 + 38
c 76 + 85
d 127 + 344
e 245 + 156

5 Totalling longer lists

a 14 + 22 + 23
b 27 + 44 + 16
c 26 + 47 + 28
d 19 + 28 + 17 + 29
e 57 + 66 + 38

6 Split these up, to work out in your head.

a 60 + 23
b 75 + 14
c 54 + 32
d 73 + 24
e 101 + 43

Subtraction

- *Understanding how subtraction relates to addition*
- *Keeping a clear head when subtraction looks complicated*
- *Mastering quick methods to do in your head*

As darkness is to light, and sour is to sweet, so subtraction is to addition. As we shall see in this chapter, this relationship between adding and subtracting is useful for understanding and calculating subtraction-based problems. If you have 7 carrots, and you add 3, and then you take away 3, you are left exactly where you started, with 7. So subtraction and addition really do cancel each other out.

Getting started with subtraction

Subtraction is also known as *taking away*, for good reason. If you have 17 cats, of which 9 are Siamese, then the number of non-Siamese cats is given by taking away the number of Siamese from the total number, that is, by subtracting 9 from 17.

Now, there is one important theoretical way that subtraction differs from addition: when we calculate 17 + 26, the answer is the same as for 26 + 17. Swapping the order of the numbers does not make any difference to the answer. But, with subtraction, this is no longer true: 26 − 17 is not the same as 17 − 26. In a later chapter we will look at the concept of *negative numbers* which give meaning to expressions such as 17 − 26. In this chapter, we will stick to the more familiar terrain of taking smaller numbers away from larger ones. (As it happens, extending these ideas into the world of negative numbers is simple: while 26 − 17 is 9, reversing the order gives 17 − 26, which comes out as −9. It is just a matter of changing the sign of the answer. But we shall steer clear of this for the rest of this chapter.)

The techniques for subtraction mirror the techniques for addition, with just a little adjustment needed. And, as with addition, the first step is to get comfortable subtracting small numbers in your head.

As ever, if you feel you could do with more practice, then set yourself your own challenges in batches of five, starting as slowly as you like, and aiming to build up speed and confidence gradually.

HAVE A GO AT QUIZ 1.

Longer subtraction

Now we move on to numbers which are more than just one digit long. These larger calculations can be set up in a very similar way to addition as this chapter's golden rule tells us.

The first thing to do is to align the two columns one above the other, making sure that units are aligned with units, tens with tens and so on. Then the basic idea is just to subtract the lower number in each column from the upper number. So to calculate 35 − 21 we would write this:

	TENS	UNITS
	3	5
−	2	1
=	1	4

EASY? THEN PRACTISE BY DOING QUIZ 2!

Taking larger from smaller: borrowing

What can go wrong with the procedure in the last section? Well, we might face a situation like this:

	TENS	UNITS
	5	6
−	2	7
=		

The first step is to attack the units column. But this seems to require taking 7 from 6, which cannot be done (at least not without venturing into negative numbers, which we are avoiding in this chapter). So what happens next? When we were adding, we had to *carry* digits between columns. In subtraction, the opposite of carrying is *borrowing*. It works like this: we may not be able to take 7 from 6, but we can certainly take 7 from 16. The way forward, therefore, is to rewrite the same problem like this:

	TENS	UNITS
	4	16
−	2	7
=		

Notice that the new top row 'forty-sixteen' is just a different way of writing the old top row 'fifty-six'. With this done, the old procedure of working out each column individually, starting with the units, works exactly as before.

What went on in that rewriting of the top row? We want to speed the process up. Essentially, one ten was 'borrowed' from the tens column (reducing the 5 there to 4) and moved to the units column, to change the 6 there to 16. Usually, when writing out these sort of calculations, we would not bother to write a little 1 changing the six to sixteen, since this can be done in your head. But if it helps you to pencil in the extra 1, then do it! It is usual, however, to change the 5 to 4 in the tens column. To take another example, if we are faced with 94 − 36, the way to write it out is like this:

	8	
	~~9~~	¹4
−	3	6
=	5	8

WHAT'S GOING ON HERE? TEST YOURSELF WITH QUIZ 3.

Subtraction with splitting

This column-based method is very reliable and efficient. But, just as we saw in the case of addition, it is not ideal when you want to calculate in your head, instead of on paper. The first purely mental technique we looked at for adding was *splitting numbers up*: to add 32 to 75, we split 32 up into 30 and 2, and then added these on separately, first 75 + 30 = 105, and then 105 + 2 = 107.

This approach works just as well with subtraction. (You might want to remind yourself of how it worked for adding before continuing.)

In the context of subtraction, it is always the number being taken away that gets split up. Suppose I know that there are 75 people in my office, of whom 32 are men. I want to know how many women there are. The calculation we need to work out is 75 − 32. The technique again involves splitting the 32 up into 30 and 2. So first we take away 30 from 75, to get 45, and then subtract the final 2, to leave the final answer of 43 women. The aim is to complete the subtraction by splitting the numbers up, without writing anything down. But, for practice, you might want to write down the intermediate step, that is, 45 in the above example.

TRY QUIZ 4. CAN YOU WORK IT OUT IN YOUR HEAD ?

Rounding up and adding on

Another mental trick we learnt for adding was *rounding up and cutting down*. This works just as well for subtraction. The only thing to watch out for is whether the numbers should be going up or down.

For example, to calculate 80 − 29, it might be convenient to round 29 up to 30. This gives us 50. It is in the final step that we need to take care. Instead of cutting the answer down by 1 (as we did when adding), this time we have subtracted 1 too many. So we have to add 1 back on, to arrive at a final answer of 51.

GIVE IT A GO YOURSELF WITH THE FINAL QUIZ, NUMBER 5.

Sum up *Subtraction is the opposite of addition. Once you know how to do one, it is just as easy to do the other!*

Quizzes

1 Getting started

a 9 − 6
b 8 − 4
c 12 − 5
d 17 − 9
e 16 − 8

2 Write in columns and subtract.

a 54 − 33
b 89 − 61
c 748 − 318
d 6,849 − 4,011
e 19,862 − 17,722

3 Get borrowing!

a 72 − 18
b 56 − 39
c 81 − 47
d 178 − 159
e 218 − 119

4 Split these up, to work out in your head.

a 60 − 23
b 75 − 14
c 54 − 32
d 73 − 24
e 101 − 43

5 Work out in your head, rounding up and adding on.

a 67 − 29
b 73 − 18
c 64 − 38
d 87 − 49
e 110 − 68

Multiplication

- *Remembering your times tables*
- *Managing long multiplication*
- *Learning some tricks of the trade*

What is multiplication? At the most basic level, it is nothing more than repeated addition. If you have five plates, each holding four biscuits, then the total number of biscuits is worked out by* adding *the numbers on each plate. So 5×4 is shorthand for five 4s being added together: $4 + 4 + 4 + 4 + 4$.

This gives us our first way to calculate the answer: as long as we can add 4 to a number, we can work out 5×4 by repeatedly adding 4: 4, 8, 12, 16, 20. The fifth number (20) corresponds to the final plate added to the biscuit collection, and so this is the answer.

We will see some slicker techniques shortly, but the perspective of repeated addition is always worth holding in the back of your mind. It also explains another word which is commonly used to describe multiplication: 'times'. The number 5×4 is the final result after 4 has been added 5 *times*.

Multiplication is usually denoted by the times symbol, $\times$. If you are working on a computer, though, often an asterisk $*$ will play that role (this was originally to prevent the times sign getting muddled up with the letter x). When we get to more advanced algebra later, we will meet other ways of writing multiplication, such as $4y$ or $4 \cdot y$.

As with addition (but not subtraction or division), the order of the numbers does not matter. So $5 \times 4 = 4 \times 5$, but the reason for this may not be completely obvious. To see why this is true, we can arrange the biscuits in a rectangular array as shown.

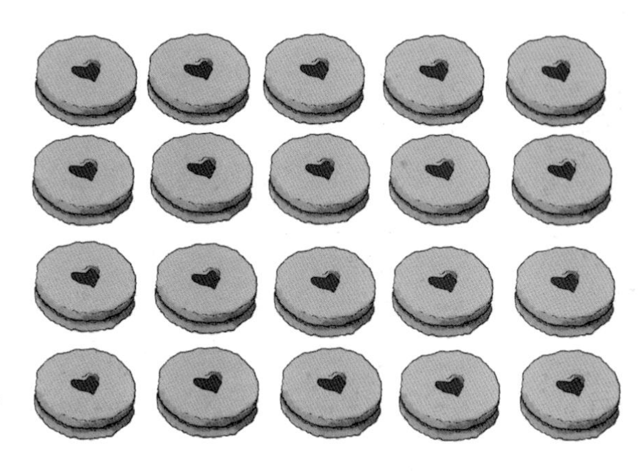

We can view this either as five columns, each containing four biscuits, giving a total of 5×4, or alternatively as four rows, each containing five biscuits, meaning that the total is 4×5. Of course this argument extends to any two numbers, meaning that for any two numbers, call them a and b, $a \times b = b \times a$.

Times tables

The trouble with the 'repeated addition' approach is that it is not practical for large numbers. To calculate 33×24 we would have to add 24s together 33 times. Most people have better ways of spending their time!

As with addition and subtraction, the key to more complex multiplication is to get to grips with the smallest numbers: 1 to 9. What this boils down to is *times tables*. For anyone hoping for an escape route, I am sorry to say that there is none! But there are some ways by which the pain can be reduced.

So here are some tips for mastering times tables:

- Firstly, remember the rule we saw above, that $a \times b = b \times a$. Once you know 6×7 you also know 7×6!
- The two times table is just *doubling*, or adding the number to itself. So $2 \times 6 = 12$ because $6 + 6 = 12$.
- The four times table means *doubling twice*. So $4 \times 6 = 24$, because $6 + 6 = 12$ and $12 + 12 = 24$.
- The five times table has a simple rule: to multiply any number (such as 7) by 5, first multiply it by 10 (to get 70) and then halve the result (35).
- The nine times table also has a nice rule. Let's look at it: $2 \times 9 = 18$, $3 \times 9 = 27$, $4 \times 9 = 36$, etc. There are two things to notice here. Firstly, all the answers have the property that their two digits add up to 9: $1 + 8 = 9$, $2 + 7 = 9$, and so on. What is more, the first digit of the answer is always 1 less than the number being multiplied by 9. So $2 \times 9 = 18$ begins with a 1, $3 \times 9 = 27$ begins with a 2, $4 \times 9 = 36$ begins with a 3, and so on. Putting these together gives us our rule: To multiply a single-digit number (such as 7) by 9, first reduce the number by 1 (to get 6). That is the first digit of the answer. The second digit is the difference between 9 and the digit we have just worked out (in this case, $9 - 6 = 3$). Putting these together, the answer is 63.

The rules so far together cover a lot, but not everything. The first things to be missed out are these four from the three times table:

$3 \times 3 = 9$ $\quad$ $3 \times 6 = 18$ $\quad$ $3 \times 7 = 21$ $\quad$ $3 \times 8 = 24$

It is also worth memorizing the *square numbers* separately, that is, numbers multiplied by themselves (see *Powers*). Some of these are covered by the rules so far. The remaining ones are:

$$6 \times 6 = 36 \qquad 7 \times 7 = 49 \qquad 8 \times 8 = 64$$

Finally we get to the trickiest ones! These are the three multiplications that people get wrong more than any others. It is definitely worth taking some time to remember them:

$$6 \times 7 = 42 \qquad 6 \times 8 = 48 \qquad 7 \times 8 = 56$$

NOW HAVE A GO AT QUIZ 1

Long multiplication

Even the most hard-working student can only learn times tables up to a certain limit. These days, the maximum is usually ten, which seems a sensible place to draw the line, and is the approach I've adopted here. When I was at school, we learnt them up to 12. The more ambitious might want to push on, memorizing times tables up to 20.

Wherever you draw the line, to tackle multiplication beyond this maximum, we need a new technique. It is time to put times tables to work!

Suppose we are asked to calculate 23 × 3. Unless we have learnt our three times table up to 23 (or our 23 times table up to 3), we need a new approach. One option is to break multiplication down into repeated addition: 23 + 23 + 23. But in the long run, a better method is to set up the calculation in vertical columns:

	2	3
×		3

To complete this, we multiply each digit of the upper number by 3, and write it in the same column below the line. As long as we know our two and three times tables, this is straightforward:

	2	3
×		3
	6	9

To calculate 41 × 4, we proceed exactly as before:

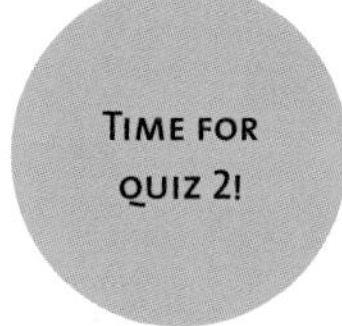

		4	1
×			4
	1	6	4

This time, the tens column produces a result of 16, and we have finished.

Carrying

Just as for addition, the moment that multiplication seems to become more complex is when the columns start interfering with each other, and the dreaded 'carrying' becomes involved again.

Well, as I hope became clear in the addition chapter, carrying is not as confusing as you might think. In fact we have already seen some carrying in this chapter. Above, when we calculate 41 × 4, the tens column ended up with 16 in it. Of course this is too many, so it was reduced to 6, and the 1 was *carried* to the hundreds column, though we may not have noticed it happening.

To take another example, let us say we want to calculate 16 × 3. If we just follow the rules above of multiplying each column separately, it comes out as follows:

	TENS	UNITS
	1	6
×		3
=	3	18

This leaves us with the correct answer, but expressed in an unusual way: thirty-eighteen. So what is that? Thinking about it, the answer must be 48.

What happens here is that the extra 1 ten from the units column gets *added* to the 3 in the tens column.

As with addition, it is usual to write the carried digits at the top as we go along. The crucial point to remember is:

Carried digits get added *(not multiplied), to the next column, after that column's multiplication has been completed.*

So, when written out, the above calculation would look like this:

	1	
	1	6
×		3
=	4	8

The 4 comes from the fact that three times 1, plus the carried 1, is 4.

Here is another example:

	2	6
×		7
=		

We begin with the units column, where $6 \times 7 = 42$. So we write down the 2 and carry the 4 to the next column:

	4	
	2	6
×		7
=		2

Next, we tackle the tens column, where $7 \times 2 = 14$, and then we add on the carried 4 to get 18:

		4	
		2	6
×			7
=	1	8	2

(Technically, the final step involved writing down 8, and carrying 1 to the hundreds column, where there is nothing else.)

Numbers march left

Which is the easiest times table? Apart from the completely trivial one times table, the answer is the ten times table. Multiplying by 10 is simple: you just have to copy the original number down, and then stick a zero at the end. So $10 \times 72 = 720$.

To say the same thing in a different way: when writing the number in columns of units, tens and hundreds, multiplying by 10 amounts to the digits of the number each taking a step to the left. So the units move to the tens column, the tens move to the hundreds column, and so on:

THOUSANDS	HUNDREDS	TENS	UNITS
0	0	7	2

→ ×10

THOUSANDS	HUNDREDS	TENS	UNITS
0	7	2	0

As always, any apparently 'empty' columns actually have a 0 in them, which is where the extra zero on the end comes from. This perspective, of the digits stepping left when multiplied by 10, is the best one for multiplication.

Another way to think of the same thing, is that in multiplying 72 by 10, we begin at the units column, with 2×10, which would give 20, but this means 0 in the units column, with 2 being *carried* to the tens column. In the same way, the 7 is carried from the tens to the hundreds column. This leftwards step, then, is nothing more than each digit being carried, without change, straight to the next column to their left.

With this in mind, multiplying by 20 or 70 becomes as easy as multiplying by 2 or 7. So $9 \times 20 = 180$, just because $9 \times 2 = 18$, and then the digits take a step to the left.

This technique combines well with the previous section. When faced with a calculation such as 53×30, we proceed exactly as for 53×3, but placing a 0 in the units column, and shifting each subsequent digit one column to the left:

It's time for a go at Quiz 3.

			5	3
×			3	0
	1	5	9	0

Putting it all together

We nearly have the techniques in place to multiply any two numbers. All that remains is to bring it all together. The critical insight at this stage is this: multiplying some number, say 74, by 52 is the same as multiplying it by 50, and separately multiplying it by 2, and then *adding* together the two answers. Remember this chapter's golden rule!

Why should this be? Suppose I am the door-keeper at a concert. The entry charge is 52 pence. To make life easy, let's suppose that everyone pays with a 50p coin and two 1p coins. If 74 people come in, then how much money have I received? The answer, of course, is the number of customers times the price: 74×52 pence. But I decide to work it out differently, and calculate the total I have received in 50p coins (74×50), and then *add* that to the amount I have received in 1p coins (74×2). Of course the answer should be the same, that is to say: $74 \times 52 = 74 \times 50 + 74 \times 2$.

The grid method

We can push this line of thought further. By exactly the same reasoning, it is also true that $74 \times 50 = 70 \times 50 + 4 \times 50$ and similarly that $74 \times 2 = 70 \times 2 + 4 \times 2$. (Just alter the numbers in the concert example!) This provides us with a way to calculate the answer to 74×52, known as the *grid method*. We work inside a grid, with one of the two numbers to be multiplied going along the top, and the other along the left-hand side. Then each of the two is split up into their tens and units components:

×		52	
		50	2
74	70		
	4		

Inside the grid, we then perform the resulting four multiplications:

×		52	
		50	2
74	70	3500	140
	4	200	8

The final stage is to *add* these four new numbers together, to arrive at the final answer: $3500 + 200 + 140 + 8 = 3848$.

The grid method easily extends to three-digit numbers. But it becomes quite time-consuming, as we have to perform nine separate calculations. For instance, to calculate 136×495 we split it up as follows:

IF YOU THINK YOU CAN MANAGE THAT, TRY QUIZ 4.

×		136		
		100	30	6
495	400			
	90			
	5			

All that remains is to fill in the gaps, and add them up.

The column method

I think the grid method for multiplication is an excellent way to get used to multiplying larger numbers. So, if you are unsure of your foothold on this sort of terrain, my suggestion is to persevere with the grid method until you get comfortable with it.

Once you are used to the grid method, however, there is another step you can take: the column method. This has the advantage of taking up less space on the page, and less time, as it needs a much smaller number of individual calculations.

Essentially the idea is to split up *one* of the two numbers into hundreds, tens, and units, as occurs in the grid method, but not the other. This amounts to calculating each row of the grid in one go. (With three-digit numbers, this reduces the list of numbers to be added from nine to three.)

As its name suggests, we are back to working in columns instead of grids. It works like this: to calculate 56×42 write the two numbers in columns.

	5	6
×	4	2

Next, ignore the '4', and simply multiply 56 by 2, by the usual method of 'carrying':

		1	
		5	6
×		4	2
	1	1	2

Then we swap: ignore the 2 in the 42 (and the new 112), and this time multiply 56 by 40. Remember that this entails multiplying by 4, and shifting the answer one step to the left:

			2	
			5	6
×			4	2
		1	1	2
	2	2	4	0

The final stage is to add the two bottom lines together:

			5	6
×			4	2
		1	1	2
+	2	2	4	0
	2	3	5	2

IT'S TIME TO TAKE ON THE FINAL QUIZ, NUMBER 5.

Sum up *Build up multiplication step by step, starting with repeated addition, until long multiplication is easy!*

Quizzes

1 A times table test!

a 2×8
b 5×6
c 6×9
d 7×7
e 7×8

2 In columns

a 34×2
b 22×4
c 31×3
d 64×2
e 41×5

3 March to the left

a 44×20
b 23×30
c 12×40
d 63×30
e 71×50

4 Multiplication in a grid

a 34×21
b 45×34
c 62×45
d 71×123
e 254×216

5 Long multiplication in columns

a 76×12
b 61×34
c 57×29
d 152×73
e 313×84

Division

- *Using times tables backwards*
- *Remembering long division*
- *Understanding chunking*

Just as subtraction is the opposite of addition, so division is the opposite multiplication. More precisely, 24 ÷ 6 is the number of times that 6 fits into 24. We could rephrase the question as '6 × [?] = 24'; by which number do we need to multiply 6 to get 24? What is this useful for? Well, suppose I want to share a packet of 24 sweets among 6 salivating children. If each child is to get the same number of sweets (seems a good idea – to avoid an almighty argument) then that number must be 24 ÷ 6.

The usual symbol for division is '÷', but computers often display it as '/'. Another way of representing division is as a fraction, so '24 ÷ 6', '24/6' and '$\frac{24}{6}$' all have exactly the same meaning.

Getting started with division

BEGIN WITH SOME PRACTICE. TRY QUIZ 1.

As usual, the starting point for division is to get used to working with the small numbers, 1 to 9. In particular it is very useful to be able to work backwards from the times tables, and to be able to answer questions like this: 6 × [?] = 42. (This is the same as calculating 42 ÷ 6.)

When things don't fit: remainders

When we are doing division with whole numbers, something rather awkward can happen, something that we didn't see with addition, subtraction or multiplication. In the case of addition, for example, if you start off with two whole numbers, then when you add them together, you will produce another whole number. But with division, this can go wrong. If we try to work out 7 ÷ 3, for example, we seem to get stuck. If we know our three times table, then we know that 7 isn't in it: the table jumps from 2 × 3 = 6 to 3 × 3 = 9. So what can we do?

Let's go back to the example of dividing up sweets between children. Suppose we have 7 sweets to divide between 3 children. To avoid a fight, we want each child to get the same number of sweets. How many can they each have? With a little reflection, the answer is 2. That leaves 1 left over, which we can put back in the bag (or eat ourselves). We can say that 7 divided by 3 is '2 with remainder 1'. We write that as:

$$7 \div 3 = 2 \text{ r } 1$$

for short. Questions like this are a tougher test of your times tables! This is how to tackle them.

- If we want to calculate $29 \div 6$, the first thing to do is to go through the six times table to find the last number in that list which is smaller than (or equal to) 29. With a little reflection, we see that number is 24.
- The next question is: $6 \times \boxed{?} = 24$? The answer is 4. So $29 \div 6$ is equal to 4, with some remainder.
- The final step is to find out what that remainder is: it is the difference between 29 and 24, which is 5. So the final answer is:

$$29 \div 6 = 4 \text{ r } 5$$

Fractions

Sometimes it is best to leave the answer to a division question as a remainder. But there are other options. To go back to the example above, where 7 sweets were divided between 3 children, we had an answer of $7 \div 3 = 2 \text{ r } 1$. One way to deal with the 1 remaining sweet is to chop it into thirds, and give each child one third. In total then, each child will have received $2\frac{1}{3}$ sweets, so $7 \div 3 = 2\frac{1}{3}$.

It is not hard to move between the language of remainders and fractions:

- Once we have arrived at 2 r 1, the main part of the answer (that's 2) remains the same.

- Then the remainder (1) gets put on top of a fraction, with the number we divided by (3) on the bottom, to give $2\frac{1}{3}$.

So, to take another example, having worked out $29 \div 6 = 4 \text{ r } 5$, we can express this as a fraction as $4\frac{5}{6}$. It is dealing with remainders which gives division its unique flavour.

REMAINDERS AND FRACTIONS. TRY QUIZ 2.

Chunking

The word 'chunking' is a fairly new addition to the mathematical lexicon, the sort of thing that might make traditional mathematics teachers raise their eyebrows. All the same, many schools around the world teach this method today. So what is chunking all about?

Actually, far from being something fancy and modern, chunking is an ancient and very direct approach to division problems involving larger numbers. It is just the *word* that is new!

Suppose we want to divide 253 by 11. The idea is to try to fit bunches of 11 inside 253, thereby breaking it up into manageable chunks. So the smaller number (11) comes in *bunches*, and the larger number (253) gets broken down into *chunks*. Got that?

Now, a bunch of ten 11s amounts to 110, and this certainly fits inside 253. In fact, it can fit inside twice, since twice 110 is 220 (but three bunches comes to 330 which is too big).

So we have broken up 253 into two chunks of 110, which with have been dealt with. The leftover is 253 − 220 which is 33. To continue, we want to fit more 11s into this final chunk. Well, 11 can fit into 33 three times. All in all then, we fitted 11 into 253 twenty times and then a further three times. So $253 \div 11 = 23$.

With chunking the key is to start by fitting in the largest bunch of 11s (or whatever the smaller number is) that you can, whether that is bunches of ten or a hundred. Doing this reduces the size of the leftover chunk, making the remaining calculation easier.

You may find it helpful to make notes as you work, to keep track of the chunks that have been dealt with, and the size of the leftover chunk.

TRY THIS YOURSELF IN QUIZ 3.

Short division

What happens when the numbers involved are larger? Suppose we are faced with a calculation like $693 \div 3$. Chunking is one option, but when the numbers are larger, it's worth knowing a careful written method.

Division is set out in a different way from the column approach of addition, subtraction and multiplication:

$$3 \overline{\smash{)}\ 6 \quad 9 \quad 3}$$

One reason for this change is that when doing addition and multiplication we work from the right (from units to tens to hundreds). In division, we work from the left, starting with the hundreds. The reason for this swap will become apparent soon!

For now, the way to approach calculations such as the above is to start with the hundreds column of the number inside the 'box' (in this case, 693, known in the jargon as the 'dividend'), and ask how many times 3 (the 'divisor') fits into it. That is to say, we begin by calculating 6 ÷ 3. The answer of course is 2, so this is written above the 6, like this:

$$\begin{array}{r|lll} & 2 & & \\ \hline 3 & 6 & 9 & 3 \end{array}$$

With this done, we move to the next step, which is to do the same thing for the tens column, and then the units. After all this, the final answer will be found written on the top of the 'box':

$$\begin{array}{r|lll} & 2 & 3 & 1 \\ \hline 3 & 6 & 9 & 3 \end{array}$$

One thing to remember is that 0 divided by any other number is still 0. So if we are working out 804 ÷ 4, when we reach the tens column, we have to calculate 0 ÷ 4. This is 0. So working it through exactly as we did above, we get:

$$\begin{array}{r|lll} & 2 & 0 & 1 \\ \hline 4 & 8 & 0 & 4 \end{array}$$

IF THAT ALL SEEMS OK, THEN HAVE A GO AT QUIZ 4!

Remainders go to work: carrying

As you might have feared, things do not always go quite as smoothly as the last section suggests. What might go wrong?

Suppose a group of 5 friends group together to buy an old car for £350. How much does each of them have to pay? The calculation we need to do is 350 ÷ 5. We can set it up as before:

$$\begin{array}{r|lll} & & & \\ \hline 5 & 3 & 5 & 0 \end{array}$$

According to the previous section, the first step is to tackle the hundreds column: 3 ÷ 5. But 5 doesn't go into 3. The five times table begins: 0, 5, 10, 15, 20, . . . with 3 nowhere to be seen. So we're stuck. What happens next?

The answer is we use our old friend 'carrying', albeit in a different guise from before. Also, remember remainders: 5 fits into 3 zero times with remainder 3. So we write a zero above the 3. But this leaves a leftover 3 in the hundreds column. This is carried to the tens column where it becomes 30. Added to the 5 that is already there, we get 35 in the tens column. That's usually written like this:

$$\begin{array}{r|ccc} & 0 & & \\ \cline{2-4} 5 & 3 & {}^{3}5 & 0 \end{array}$$

Now we can carry on as before: since 35 ÷ 5 = 7:

$$\begin{array}{r|ccc} & 0 & 7 & 0 \\ \cline{2-4} 5 & 3 & {}^{3}5 & 0 \end{array}$$

So we arrive at an answer of 70.

What happened during this new step was that we essentially split up 350 in a new way. Instead of the traditional 3 hundreds, 5 tens and 0 units, we rewrote it as 0 hundreds, 35 tens, and 0 units. With this done, the calculation could proceed exactly as before.

Let's take another example. Say 984 ÷ 4. As ever, the first thing to tackle is the hundreds column, where we face 9 ÷ 4. This is slightly different from the last example, where we had 3 ÷ 5. In that case, 5 could not fit into 3 at all; it was just too big. But this time 4 does fit into 9. The answer is 2, with a remainder of 1. This remainder gets carried to the next column. The 2 is written above the 9. That takes us this far:

$$\begin{array}{r|ccc} & 2 & & \\ \cline{2-4} 4 & 9 & {}^{1}8 & 4 \end{array}$$

The next stage is to tackle the tens column, where we have 18 ÷ 4. Once again, this doesn't fit exactly, but gives an answer of 4 with remainder 2. So the 4 gets written above the 8, and the remainder is carried to the next column:

	2	4	
4	9	$^{1}8$	$^{2}4$

TIME TO PRACTISE THESE, IN QUIZ 5.

The final step is the units column, where we have 24 ÷ 4. That is 6. So we have our final answer: 246.

Long division

There are few expressions in the English language that induce as much horror as 'long division'. In fact, it's not so bad. Long division is essentially the same thing as the short division we have just met. It's just a little bit longer.

The difference is that as the numbers involved become larger we may have to carry more than one digit at a time to the next column. So calculating the remainders becomes more cumbersome. Rather than cluttering up the division, the remainders are written underneath instead. So, if we wanted to calculate 846 ÷ 18, short division would look like this:

	0	4	7
18	8	$^{8}4$	$^{12}6$

while long division occupies a little more space:

	0	4	7
18	8	4	6
	8	4	
	−7	2	
	1	2	
	1	2	6

What is the meaning of the column of numbers underneath?

Since 18 cannot divide the 8 in the hundreds column, we carry the 8 and move on to the next column. The only difference is that we write the 84 underneath this time. Then 18 goes in to 84 four times, since $4 \times 18 = 72$, but $5 \times 18 = 90$ which is too big. So 4 is written on top, just as before, and 72 is written below 84 and then subtracted from it to find the remainder, 12.

If we were doing short division, 12 would be the number we carry to the next column and stick in front of the next digit. But because we are doing things underneath, we bring down the next digit from 846 (namely 6) and stick it on the end of the 12 to get 126. The last step is to try to divide 126 by 18. A little chunking shows that 18 fits in exactly 7 times, so 7 is written on the top, to complete the calculation.

Dare you try long division? Don't be put off by the numbers underneath: if you're not sure what you should be writing down there, try laying the whole thing out as a short division, and doing any supplementary calculations you need underneath. Remember: the working underneath is intended to help you with the calculation, not to confuse you!

IF YOU'VE GOT THE NERVE, TRY QUIZ 6!

Sum up *There are several methods for bringing division down to earth. But even long division is manageable, once you have a good grasp on remainders!*

Quizzes

1 Times tables, backwards!

a $4 \times \boxed{?} = 12$
b $5 \times \boxed{?} = 30$
c $3 \times \boxed{?} = 27$
d $8 \times \boxed{?} = 64$
e $9 \times \boxed{?} = 63$

2 Write out as remainders and as fractions.

a $11 \div 4$
b $16 \div 6$
c $24 \div 7$
d $48 \div 5$
e $59 \div 8$

3 Chunking

a $96 \div 8$
b $154 \div 7$
c $279 \div 9$
d $372 \div 6$
e $8488 \div 8$

4 Lay these out as short divisions.

a $864 \div 2$
b $770 \div 7$
c $903 \div 3$
d $8482 \div 2$
e $9036 \div 3$

5 Short division

a $605 \div 5$
b $426 \div 3$
c $917 \div 7$
d $852 \div 6$
e $992 \div 8$

6 Long division! Dare you try it?

a $294 \div 14$
b $270 \div 15$
c $589 \div 19$
d $1785 \div 17$
e $1464 \div 24$

Primes, factors and multiples

- *Understanding prime numbers and why they are so important*
- *Being able to tell when one number is divisible by another*
- *Knowing how to break a number down into its basic components*

Odd numbers, even numbers, prime numbers, composite numbers, square numbers, . . . these are just a few of the different types of numbers that mathematicians get incredibly excited about. What are all these different sorts of number? Most of these terms refer to the different ways that whole numbers are built out of others. This will become clearer when we have met the most important numbers of all: prime numbers.

Prime numbers

The definition of a *prime number* is simple: a prime number is a whole number which cannot be divided by any other whole number (except 1 and itself). So, for example, 3 is prime because the only way to write 3 as two positive whole numbers multiplied together is as 3×1 (or 1×3, which is essentially the same thing). On the other hand 4 is not prime because $4 = 2 \times 2$.

A *composite number* essentially means a 'non-prime' number, and 4 is the first example. Similarly 5 is prime, but 6 is composite. (The numbers 0 and 1 are so special that they deserve categories of their own, and are classed as neither prime nor composite.)

The first 25 primes are: 2 3 5 7 11 13 17 19 23 29 31 37 41 43 47 53 59 61 67 71 73 79 83 89 97

It was Euclid, in around 300 BC, who first proved that the list of primes goes on forever. There is no largest prime number, and so people keep finding bigger and bigger ones. It is a tough job though, as telling whether a very large number is prime or composite is hard. The largest prime known so far is 12,978,189 digits long!

GOLDEN RULE

Every whole number can be broken down into prime numbers, in only one way.

The atoms of mathematics

Why do people get so excited about prime numbers? The reason they are so important is that they are the fundamental blocks from which all other numbers are built. Although 6 is not prime, it can be broken down into primes as 3×2. Similarly 8 can be broken down as $2 \times 2 \times 2$, and 12 as $2 \times 2 \times 3$. In this sense, prime numbers are like mathematical atoms: everything else is built from them.

What is more, this chapter's golden rule says a little bit more than this. Not only can every number be broken down into primes, but there is *only one* way to do it. So once we know that $1365 = 3 \times 5 \times 7 \times 13$, for example, it follows that the only other ways to write 1365 as a product of prime numbers are reorderings of this: $5 \times 3 \times 13 \times 7$, for example. So we know automatically, without having to check, that $1365 \neq 5 \times 5 \times 5 \times 11$ (the symbol $\neq$ means 'is not equal to'). This rule goes by the grand title of *The fundamental theorem of arithmetic*.

Break some numbers down into primes in Quiz 1.

Even and odd

Even numbers are those which appear in the two times table: 2, 4, 6, 8, 10, . . . Another way to say the same thing is that even numbers are those which have 2 as a *factor*, meaning that 2 can divide into the number exactly, without leaving a remainder. Yet another way to say the same thing, is that the even numbers are the *multiples* of 2.

Odd numbers, of course, are the remaining numbers: the numbers which do not have 2 as a factor.

Factor and *multiple* are opposite terms. To say that 15 is a multiple of 3 is the same as saying that 3 is a factor of 15. Both statements mean that 3 can divide into 15 exactly, without leaving a remainder. In other words, 15 is in the three times table.

Try out these terms in Quiz 2.

Divisibility tests

It is often useful to know whether or not a large number is a multiple of a particular smaller number. For some small numbers this is so easy that we can do it without thinking:

- The multiples of 2 are exactly the even numbers, meaning all the numbers that end in 2, 4, 6, 8 or 0.
- The multiples of 5 are the numbers that end in a 5 or a 0, such as 75 and 90.
- The multiples of 10 end in 0s, such as 80, 250, 16,700.

For other small numbers there are other tests, which are slightly subtler:

- You can tell whether or not a number is a multiple of 3 by adding up its digits. If the total is a multiple of 3, then so was the original number.

So 117 is a multiple of 3, because 1 + 1 + 7 = 9, which is a multiple of 3. On the other hand 298 is not a multiple of 3, because 2 + 9 + 8 = 19.

- A number is a multiple of 6 if it passes the tests for 2 and 3. So 528 is divisible by 6, since it is even, and 5 + 2 + 8 = 15, which is divisible by 3. (Notice that the total of the digits does *not* have to be divisible by 6.)

- The test for divisibility by 9 is similar to the test for 3: add up the digits, and if the result is a multiple of 9, then so was the original number. So 819 is a multiple of 9, since 8 + 1 + 9 = 18, but 777 is not, since 7 + 7 + 7 = 21.

- You can tell whether a number is a multiple of 4 just by looking at its last two digits. If they are a multiple of 4, then so is the whole thing. So 116 is a multiple of 4, just because 16 is. Similarly 5422 is not a multiple of 4, as 22 isn't.

- The number 8 is a little awkward, and there are various possible ways forward. One is a variation on the test for divisibility by 4. (Another is to give up and use a calculator!) If the last three digits of the number are divisible by 8, then so is the original number. So 6160 is divisible by 8, since 160 is. The trouble is that telling whether a three-digit number is divisible by 8 is not something most people can do on sight.The best option is to divide the three-digit number by 2, and then apply the test for divisibility by 4. So if we want to know whether 7476 is divisible by 8, first take the last three digits (476) and then divide by 2 (238) and finally look at the last two digits of that (38). In this case that is not a multiple of 4, so the number fails the test.

- The fiddliest single-digit number is 7. There is a workable test though,and it goes like this. To test 399 for divisibility by 7, chop off the last digit (9) and double it (18). Then subtract that from the truncated number (39 − 18 = 21). If the result is divisible by 7, then so is the original number, which in this case it is. With this test we might end up with 0: for instance if we apply the test to 147, we get 14 − 14 = 0. In this situation, 0 *does* count as a multiple of 7, and so the number passes the test.

PRACTISE THESE TESTS IN QUIZ 3!

- The number 11 has a lovely test! It goes like this. Go through the digits, alternating between adding and subtracting. If the result is divisible by 11, then so is the original number. To test 9158, we go 9 − 1 + 5 − 8 = 5, which is not divisible by 11, so the test is failed. It's possible to end up with 0 again, or even negative numbers, but that's no problem. We do count 0 and −11, and −22, and so on, as multiples of 11. So 1914 is a multiple of 11 since 1 − 9 + 1 − 4 = −11 is divisible by 11.

Breaking a number down into primes

Earlier in the chapter, we said that every number can be broken down into primes, and we saw some examples. But if we are given a larger number, such as 308, how can we actually find out what its prime ingredients are? The idea is to try dividing by prime numbers in turn, using the tests we've just seen. To start with, 308 is undoubtedly even. So we can divide it by 2, this leaves 154. This is also even, so we can divide it by 2 again, to get 77. Now, this is no longer even, so we exhausted the 2s, and we move on to the next prime. We might try dividing 77 by 3, but it fails that test. It is also easy to see that 77 is not divisible by 5. So the next prime on the list is 7, and 77 is indeed divisible by 7. Dividing it by 7 leaves 11, which is itself prime. So we have finished. Collecting together all the primes that we divided by, we get: $308 = 2 \times 2 \times 7 \times 11$.

BREAK SOME LARGER NUMBERS DOWN IN QUIZ 4.

The mysteries of the primes

The prime numbers are as mysterious as they are important, even today. If you look at the sequence of prime numbers, there seems to be very little order to it. Sometimes primes come very close together, like 11 and 13, and sometimes there are larger gaps such as between 199 and 211.

There are lots of seemingly basic facts about the prime numbers that we still do not know for sure. One of these is *Goldbach's conjecture*. In 1742, Christian Goldbach noticed that every even number from 4 onwards is actually the *sum* of two prime numbers. So $4 = 2 + 2$, $6 = 3 + 3$, $8 = 3 + 5$, . . . If you can prove that Goldbach's conjecture must be true for *every* even number, then you will have outshone the mathematicians of the last two centuries. Although it has been verified up to an enormous limit (around 10^{18} – see *The power of 10* for what this means), no-one has yet managed to prove that it must be true for *all* even numbers.

TRY GOLDBACH'S CONJECTURE FOR YOURSELF IN QUIZ 5.

Sum up *To get to know a number well, you need to know which other numbers divide into it. The most important ones to check are the atoms of the mathematical world, the primes!*

Quizzes

1 Break these numbers down into primes.

a 15
b 18
c 21
d 24
e 32

2 Which are true and which are false?

a '18 is a multiple of 3'
'18 is a factor of 3'

b '246 is a multiple of 5'
'5 is a factor of 246'

c '4 is a multiple of 108'
'108 is a factor of 4'

d '114 is a multiple of 6'
'6 is a factor of 114'

e '245 is a multiple of 7'
'7 is a factor of 245'

3 Test these numbers for divisibility up to 11.

a 64
b 42
c 75
d 176
e 68

4 Break these numbers down into primes.

a 30
b 210
c 108
d 189
e 1617

5 Goldbach's conjecture! Write these even numbers as two primes added together.

a 10
b 12
c 14
d 16
e 18

Negative numbers and the number line

- *Understanding what negative numbers mean*
- *Recognizing when negative numbers are useful*
- *Knowing how to use the 'number line'*

If the idea of* negative numbers *does not come naturally to you, don't worry. You are in good company! It took mathematicians and scientists thousands of years before the concept became respectable. But if you don't have a thousand years to spare, you needn't worry either. The principle is quite simple, once the basic idea has been grasped.

Negative numbers

The story of negative numbers begins in the world of commerce, and they still demonstrate their great usefulness in trade today.

Imagine that I have set up a business, and am looking back over my accounts at the end of my first month's trade. There are three basic positions that I might be in. Firstly, if my bank account is overdrawn, that means that I am in debt. Over the month, I have spent more money than I have received. So, how much money do I actually have, at this stage? The true answer is 'less than zero'.

The second possibility is that I have broken even. If my expenditure and income have balanced each other out exactly, then the amount of money in my account is zero. I am neither in debt, nor in credit.

The third possibility is that more money has come in than I have spent. In other words, I have made a profit, and my bank account is in credit.

(Of course, this is a simplification from a business perspective, where people generally distinguish between capital investment at the start of a business, and running expenses. Nevertheless, the essential idea is, I hope, reasonable enough.)

In the past, people considered these three separate possibilities as being essentially different. But, over time, the realization dawned that the three could all be represented as different positions along a single scale. Nowadays, we call this picture the *number line*.

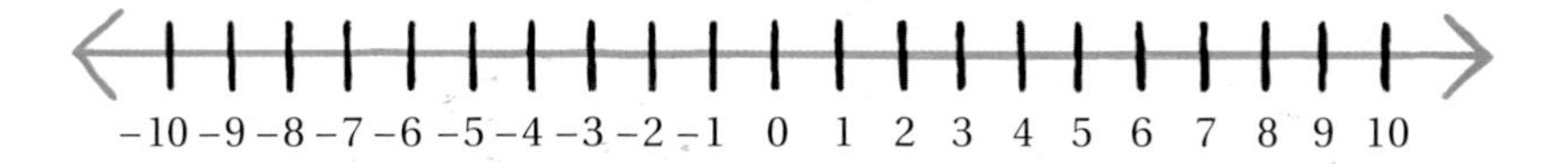

The number line

The number line is a horizontal line, with 0 in the middle. To the right of 0, the positive numbers line up in ascending order: 1, 2, 3, 4, . . . To the left of zero are the *negative numbers*, which progress leftwards: -1, -2, -3, -4, . . .

Sometimes is it convenient to put negative numbers inside brackets like this: (-1), (-2), (-3), . . . There is nothing complicated going on here; it is just to stop the $-$ signs getting muddled up when we start having other symbols around.

Notice that there is no -0. At least there is, but it is the same thing as the ordinary zero: $-0 = 0$. Every other number is different from its negative, so $-1 \neq 1$, for instance.

We might think of this number line as representing my bank account. At any moment, it is at some position along that line. If I am £15 overdrawn, I am at -15. If I am £20 in credit, I am at $+20$. (It is usual to omit the plus sign, and just write '20', but sometimes it is useful to include it for emphasis.)

For this chapter, we will be focusing on the whole numbers (positive, negative and 0). But between these are all the usual decimals and fractions, which also come in both positive and negative varieties. We shall meet these in more detail in future chapters. But if we want to find $4\frac{2}{3}$, it is $\frac{2}{3}$ of the way from 4 to 5. In the same way, $-4\frac{2}{3}$ is $\frac{2}{3}$ of the way from -4 to -5. (A possible mistake here is to position it as $\frac{2}{3}$ of the way *from* -5 *to* -4.)

HAVE A GO AT THIS YOURSELF IN QUIZ 1.

Negative numbers in the real world

For years, the principal purpose of numbers has been to count things: 3 apples, 7 children or 10 miles. So, when negative numbers first make their entrance, a natural question is: how can you have -3 apples? I hope that an answer is now plausible: having -3 apples means being *in debt* by 3 apples.

How does the number line tie in with the usual idea of addition? Well suppose you now go and pick 3 apples. But, instead of adding them to your apple larder, you pay them to the person to whom you owe 3 apples. So, after receiving 3 apples, you end up with none: $-3 + 3 = 0$. This can be shown on the number line as starting at -3, then moving three places to the right to end up at 0.

It is not just trade where negative numbers are useful. Another example is temperature. In the Celsius (or centigrade) scale, 0 is defined to be the freezing point of water. If we start at 0 degrees and gain heat, we move up into the warmer, positive temperatures. If we lose heat, we move downwards into the colder, negative numbers. A thermometer, then, is nothing more than a number line, with a tube of mercury giving our current position on it.

Moving along the number line

The number line is useful for seeing addition and subtraction at work. If I am at 7, then adding 3 is the same as taking three steps to the right along the number line: $7 + 3 = 10$. Similarly, subtracting 3 is the same as taking 3 steps to the left: $7 - 3 = 4$.

This is not exactly news. But the same principle remains true whatever the starting position. So even if we begin at a negative number such as -5, then adding 3 again means taking three steps right: $-5 + 3 = -2$. Similarly, subtracting 3 means taking 3 steps left: $-5 - 3 = -8$.

PRACTISE USING THE NUMBER LINE IN QUIZ 2.

Negative negatives

Above, we saw how to use the number line to add or subtract. But there is still something we need to make sense of: what is the relationship between *subtraction* and *negative numbers*? In a sense they are the same thing . . . but we need to know the details.

The trouble is that we seem to be using the same symbol ($-$) for two different things: firstly (as in '-3'), this symbol indicates a *position* on the number line to the left of 0, meaning a negative number; and secondly, to describe a way of combining two numbers, as in '$7 - 4$'. This second use corresponds to a *movement* leftwards along the number line.

So what is going on here? You could think of putting a minus sign in front of a number as like 'flipping it over', using 0 as a pivot. So putting a minus sign in front of 7 means 7 flips over, all the way to the far side of 0, and lands on -7. So what then is '$- -7$', or '$-(-7)$', as we might write it? Well, when you flip over -7 you get back to 7. So:

$$-(-7) = 7$$

The fact that two minus signs *cancel each other out* in this way is the key to working with negative numbers.

So when we face questions like '$9 - (-3)$', the two minus signs cancel out, to give us '$9 + 3$'. But when we have '$9 + (-3)$', there is only one minus sign, so it doesn't get cancelled out, and is the same thing as '$9 - 3$'. This then is the relationship between negative numbers and subtraction:

Subtracting 3 *from* 9 *is the same thing as adding* -3 *to* 9.

Notice that this is *not* the same as adding -9 to $+3$.

Now have a go at quiz 3.

Multiplying

So much for addition and subtraction. What about multiplying negative numbers? Well to start with, remember that multiplication is essentially repeated addition. So $4 \times (-2)$ should be the same as $(-2) + (-2) + (-2) + (-2)$, which is just $-2 - 2 - 2 - 2$, that is to say -8.

To think about this in terms of trade, if I lose £2 each day (that is to say, if I 'make $-$£2'), then after four days I have lost £8 (or 'made $-$£8'). This illustrates that when we multiply a positive number by a negative number, the answer is negative. So $-5 \times 2 = -10$ and also $5 \times -2 = -10$. Similarly, $-1 \times 4 = -4$ and $1 \times -4 = -4$, and so on.

The most confusing moment in the dealing with negative numbers is when two negative numbers are multiplied together: $(-4) \times (-2)$, for example. But we have already seen above how two minus signs cancel each other out, and here it is exactly the same again. Two negative numbers produce a *positive* result: $(-4) \times (-2) = 8$.

How does this work in terms of trade? Suppose I lose £2 per day (that is to say, I 'make $-$£2'). The question is how much will I have made or lost in -4 days time? Well, 'in -4 days time' must mean 4 days ago. And if I have been losing money at a rate of £2 per day, then 4 days ago I must have been £8 richer than I am today, which matches the result above.

We can put these rules in a little table:

MULTIPLIED BY	NEGATIVE	POSITIVE
NEGATIVE	Positive	Negative
POSITIVE	Negative	Positive

Or more concisely:

×	−	+
−	+	−
+	−	+

GOT IT? TRY QUIZ 4.

Dividing

When you have mastered negative multiplication, division is easy! All we need to do is 'multiplication backwards'. So to calculate $(-6) \div (-3)$, we have to solve $(-3) \times \boxed{?} = -6$. The two obvious possibilities are 2 and -2, but only one can be right, so which is it? Well we know that $(-3) \times (-2) = 6$, which is not what we want. But $(-3) \times 2 = -6$, exactly as we might hope. So the answer is 2.

Perhaps surprisingly, the rules for working out the sign for division are the same as for multiplication:

DIVIDED BY	NEGATIVE	POSITIVE
NEGATIVE	Positive	Negative
POSITIVE	Negative	Positive

Or more concisely:

÷	−	+
−	+	−
+	−	+

TRY USING THESE RULES IN QUIZ 5.

Sum up *A number line is a great picture of the world of numbers: positive, negative, and zero.*

Quizzes

1 Draw a number line, between −5 and +5. Mark all the whole numbers. Then add in marks for these numbers.

a $-\frac{1}{4}$ and $\frac{1}{4}$ **b** $-1\frac{1}{2}$ and $1\frac{1}{2}$

c $-2\frac{1}{3}$ and $2\frac{1}{3}$ **d** $-3\frac{3}{4}$ and $3\frac{3}{4}$

e $-4\frac{2}{5}$ and $4\frac{2}{5}$

2 Add and subtract on a number line.

a $8 + 7$ and $8 - 7$

b $3 + 3$ and $3 - 3$

c $3 + 6$ and $3 - 6$

d $-5 + 4$ and $-5 - 4$

e $-2 + 3$ and $-2 - 3$

3 Doubling back

a $5 + (-4)$ and $5 - (-4)$

b $2 + (-3)$ and $2 - (-3)$

c $0 - 5$ and $0 - (-5)$

d $-4 - 2$ and $-4 - (-2)$

e $-3 - 5$ and $-3 - (-5)$

4 Times tables go negative

a $2 \times (-3)$ and $(-2) \times (-3)$

b $4 \times (-5)$ and $(-4) \times (-5)$

c $7 \times (-3)$ and $(-7) \times (-3)$

d $8 \times (-4)$ and $(-8) \times (-4)$

e $25 \times (-4)$ and $(-25) \times (-4)$

5 Division goes negative

a $8 \div 2$ and $8 \div (-2)$

b $(-18) \div 6$ and $(-18) \div (-6)$

c $28 \div 7$ and $28 \div (-7)$

d $(-33) \div 3$ and $(-33) \div (-3)$

e $(-57) \div 19$ and $(-57) \div (-19)$

Decimals

- *Interpreting decimals such as 0.0789*
- *Understanding what happens to the decimal point during arithmetic*
- *Mastering rounding*

Not everything can be measured as whole numbers. It may not take a whole number of minutes to walk to the shop, a recipe may not require exactly a whole number of litres of milk. When we need to divide things up more finely than the whole numbers allow, there are two main approaches: fractions and decimals.

Neither method is better than the other; both are in use all the time. So it is important to be able to translate between the two. As a very rough rule of thumb, when the fraction is a simple one, it is best to use that: so we might speak of 'half an apple', or 'three quarters of a mile'. But it is not practical to talk about 'thirteen twenty-sevenths of a litre'. So when real precision is needed, I recommend decimals.

Decimals – what's the point?

With that all said, what exactly is a 'decimal'? The idea comes from the column representations of whole numbers that we have met in earlier chapters. There, we had columns for units, tens, hundreds, thousands, and so on, like this:

HUNDREDS	TENS	UNITS
7	8	4

To incorporate things smaller than units, this system gets extended. We introduce new columns for tenths of a unit, and similarly for hundredths, thousandths, and so on, like this:

HUNDREDS	TENS	UNITS	TENTHS	HUNDREDTHS
7	8	4	9	2

There is a mental adjustment we need to make when working with decimals. When we are just writing whole numbers, we know that the units column is the always the one furthest to the right. So '28' must mean this:

HUNDREDS	TENS	UNITS
0	2	8

But when we move into the realm of decimals, we have new columns to the right of the units. Now the digits alone do not make it clear where the units are, or which column is on the right. So '287' might mean:

HUNDREDS	TENS	UNITS	TENTHS	HUNDREDTHS
0	2	8	7	0

or:

HUNDREDS	TENS	UNITS	TENTHS	HUNDREDTHS
0	0	2	8	7

or many other variations on the theme. This is a disaster!

The problem is solved with a new ingredient: the 'decimal point'. This is a dot which sits to the right of the units column. It is this point which anchors the columns, allowing us to tell which is which. So '28.7' means:

HUNDREDS	TENS	UNITS	TENTHS	HUNDREDTHS
0	2	8	7	0

while '2.87' means:

HUNDREDS	TENS	UNITS	TENTHS	HUNDREDTHS
0	0	2	8	7

One consequence is that whole numbers directly translate into decimals with the addition of 0s in the columns for tenths, hundreds, thousandths, etc. So '34.0', '34.00', '34.000', and so on, all mean exactly the same as '34'.

(Some countries and languages use a 'decimal comma' instead of a decimal point, but it serves exactly the same purpose.)

Decimal arithmetic: addition and subtraction

The great thing about decimals is that the old column-based methods of arithmetic transfer straight over to this new context. So to calculate 3.3 + 5.8, we set it up in columns exactly as we did for addition of whole numbers, just making sure to line up the decimal points of all the numbers involved:

	3	·	3
+	5	·	8
=		·	

We proceed as before, making sure to start with the rightmost column, and then carrying 1 to the next column as needed:

If you're happy adding whole numbers, try quiz 1.

	1		
	3	·	3
+	5	·	8
=	9	·	1

(If you're not yet fully comfortable adding whole numbers, you might want to revisit the chapter on addition.)

What goes for addition is equally true of subtraction. To calculate 6.2 − 2.4, we set it up in columns like this:

	6	·	2
−	2	·	4
=		·	

Again the method is identical to that for whole numbers, presented in the chapter on subtraction. So we begin with the rightmost column, and borrow 1s as necessary:

Try this yourself in quiz 2.

	5		
	~~6~~	·	2
−	2	·	4
=	3	·	8

Multiplication: moving rightwards

Like addition and subtraction, multiplcation also translates easily to the decimal context. The main thing to watch out for is the position of the decimal point, or more accurately the positions of the digits *relative* to

the decimal point. It is a good habit to think of the decimal point as being fixed and immovable, while the digits around it shuffle leftwards or rightwards.

To see what we mean by the 'position of the digits relative to the decimal point', let's look at the calculation 3×2, but with these digits in different places relative to the decimal point. Let's start with 3.0×2.0. There are no surprises here, since this is nothing more than 3×2, which we know to be 6. Another easy one is $3.0 \times 20.0 = 60$, which we might calculate as $20 + 20 + 20 = 60$. Alternatively, we could first work out 3×2, and then shunt the answer one column to the left, filling in the empty column with a zero, again giving 60. (This is the method presented in the chapter on multiplication.)

Next, what is 3.0×0.2? The answer is 0.6. We can see this easily, because $0.2 + 0.2 + 0.2 = 0.6$.

But how does this fit in with the column depiction of the calculation?

The golden rule here is this: just as multiplying by 10 moves the digits one step to the *left*, so multiplying by 0.1 moves them one step to the *right*. Why should this be? One answer is that multiplying by 0.1 (that is to say $\frac{1}{10}$) is the same thing as *dividing* by 10.

To calculate 3×20, the first step was to realize that $3 \times 2 = 6$, and the second was to shift everything one column to the left, filling in any empty columns with zeros, to arrive at 60. Calculating 3×0.2 is almost the same: first we calculate 3×2, and then shunt everything one column to the *right*, again filling any empty columns with zeros, to arrive at 0.6.

The same line of thought works when we look at 0.3×0.2. As before we first work out 0.3×2, which is 0.6, and then shunt everything one step to the right, to arrive at an answer of 0.06.

IF THAT SEEMS TO MAKE SENSE, HAVE A GO AT QUIZ 3.

Multiplication: putting it together

With the golden rule in place, more complex multiplication can be put together just as for whole numbers. The pattern follows the column method for whole number multiplication exactly (so make sure you are comfortable with that before proceeding with this). To calculate 21.3×3.2, we would set it out like this:

	2	1	·	3
×		3	·	2
			·	

We start on the right, by multiplying the whole top row by the rightmost digit on the bottom row. In this case, that means multiplying it by 2, and then moving everything one step to the right:

	2	1	·	3	
×		3	·	2	
		4	·	2	6

Then we move to the next digit of the second number, in this case 3, and again multiply the top number by that. This time no shunting, left or right, is required because 3 is in the units column:

	2	1	·	3	
×		3	·	2	
		4	·	2	6
	6	3	·	9	

Finally, we add up the two numbers below the line to arrive at the final answer:

	2	1	·	3	
×		3	·	2	
		[1]4	·	2	6
	6	3	·	9	
	6	8	·	1	6

Now have a try yourself, in Quiz 4.

Notice how the decimal points are kept in line throughout the calculation. This is good practice, as it means all the columns marry up correctly.

Rounding

We use decimals when we want a more accurate description than whole numbers can provide. But we may not always want the laser-sharp accuracy that decimals of unlimited length provide. This might sound strange, but think of it this way: do you need to know the quantity of butter to use in a cake, down to millionths of a gram? It is useful for me to know my weight to the nearest kilogram, or even in exceptional circumstances to the nearest gram, but never to the nearest nanogram (that is, 0.000000001g, or one billionth of a gram).

Very often we want to *round* decimals to some chosen level of accuracy. If I ask your height to the nearest centimetre, you might give me an answer of 1.64 metres. In doing so, you have rounded your answer to *two decimal places*.

Rounding is something we shall use throughout this book. So how does it work? We begin by specifying a level of accuracy, usually as a number of decimal places. The *decimal places* are the columns to the right of the decimal point: those representing tenths, hundredths, thousandths, etc.

Suppose we decide on one decimal place as a suitable level of accuracy for measuring the weight of a helping of dog-food. This means we will want to round our answer to the nearest tenth of a kilogram. The scale reads 3.734kg. So, when we round it, the answer is 3.7kg.

It is tempting to think that rounding is simply a matter of chopping the number off at a suitable point. Actually there is a little more to it than this – but just a little. For example, suppose we want to round 3.699 to 1 decimal place. Chopping the end off would give 3.6, but in fact 3.699 is much closer to 3.7. It is just 0.01 away from 3.7, whereas it is 0.099 away from 3.6. So 3.7 is the correct answer here.

Now, what if we wanted to round 5.46 to 1 decimal place? The two candidates are 5.4 and 5.5. But 5.46 is 0.06 away from 5.4, and just 0.04 away from 5.5, so 5.5 is the right answer. On the other hand, if we start with 2.13, then the two candidates are 2.1 (0.03 away) and 2.2 (0.07 away). So, this time, the answer is 2.1. Thinking about different examples like these produces the following rule:

- To round a number to a certain number decimal places, chop it off after those digits.

- Then look at the first digit of the tail (the part that was chopped off).
- If that digit is between 0 and 4, then leave the answer as it stands: just the original number, truncated.
- But if the first digit of the tail is between 5 and 9, then the final digit of the truncated number should be increased by 1.

This rule is, admittedly, a bit of a mouthful. But the procedure itself is not really very difficult, with a little practice. The trickiest case is when we have an example like 19.981, to be rounded to 1 decimal place. We begin by chopping off after 19.9. Then, the first digit of the tail is 8. So we know we have to increase 19.9 by 0.1 to 20.0. In this case the rounded answer doesn't actually look anything like the original number, which might be confusing at first.

Incidentally, when rounding to one decimal place, it is good practice always to include that decimal place, even if it contains a zero. So in the above example, we would leave the answer as '20.0' rather than abbreviating it to '20'. The reason for this is that '20.0' communicates the level of accuracy to which you are working, namely one decimal place.

HAVE A GO AT QUIZ 5!

There will be plenty more practice in rounding later in the book.

Sum up *If you can do arithmetic with whole numbers in columns, then it is only a small step to extend the technique to decimals.*

Quizzes

1 Set out in columns and add

a 3.2 + 2.3
b 6.4 + 6.7
c 12.31 + 3.19
d 6.78 + 3.33
e 0.00608 + 0.00503

2 Decimal subtraction (columns again)

a 7.9 − 3.6
b 6.43 − 2.31
c 9.6 − 1.7
d 7.67 − 3.48
e 19.72 − 9.89

3 Digits move right

a 0.2 × 4
b 0.2 × 0.4
c 0.5 × 7
d 0.5 × 0.7
e 0.5 × 0.07

4 Full-blown multiplication

a 2.3 × 1.7
b 6.2 × 5.2
c 3.4 × 2.9
d 25.7 × 6.8
e 5.72 × 7.9

5 Round these.

a 5.3497 to one decimal place
b 0.16408 to two decimal places
c 0.16408 to one decimal place
d 9.981123 to one decimal place
e 0.719601 to three decimal places

Fractions

- *Interpreting fractions*
- *Recognizing when two fractions are the same*
- *Understanding top-heavy fractions*
- *Translating between fractions and decimals*

Working with fractions ought to be easy. All it involves is cutting up cakes into suitably sized slices, and then counting the pieces. But somehow this simple idea can turn into a nightmare of 'common denominators' and 'lowest common multiples'.

There is one key thing to understand to be able to work with fractions. It is that any fraction, such as $\frac{1}{2}$, can be rewritten in different ways, for example as $\frac{2}{4}$ or $\frac{5}{10}$. The crucial question is this: how can we tell when two fractions are really the same? Might $\frac{1}{2}$ also the same thing as $\frac{13}{27}$, for example?

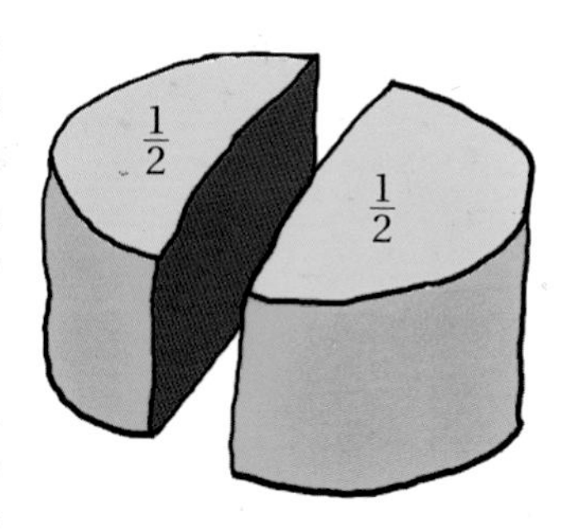

An old reliable method is to start with two identical cakes and divide up each according to the two fractions we are thinking about. If the two resulting helpings are the same, then so are the two fractions. So, if we divide one strawberry pavlova into eighths, for example, and serve up four of them, and divide another pavlova into two halves, and serve up one of them, then the two helpings are indeed the same. So, $\frac{4}{8} = \frac{1}{2}$. But if we divide a cake into twenty-sevenths, and dish out 13 of them, we do not quite get a helping equal to half a cake. So $\frac{13}{27} \neq \frac{1}{2}$.

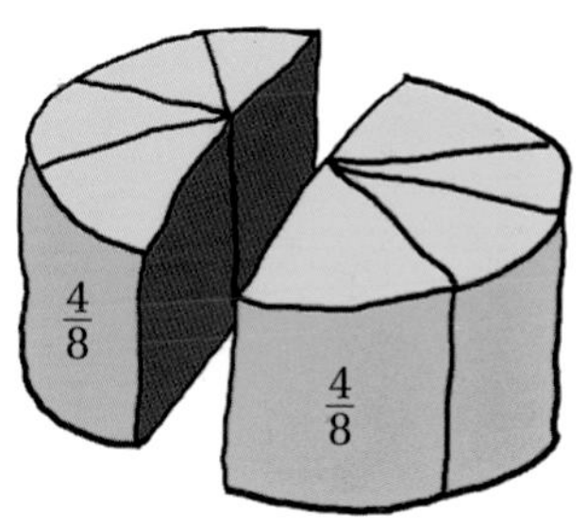

What we need is a way to get this information directly from the numbers, without having to bake any cakes. As it happens, it is not too hard in this particular case: four eighths is the same as one half, because 4 is half of 8. On the other hand, 13 is not half of 27. This is fine, but some numbers are less easy to manipulate than $\frac{1}{2}$. What this is hinting towards is the golden rule which tells us how to change a fraction's appearance, while keeping its value the same.

IF YOU HAVE GRASPED THIS, HAVE GO AT QUIZ 1.

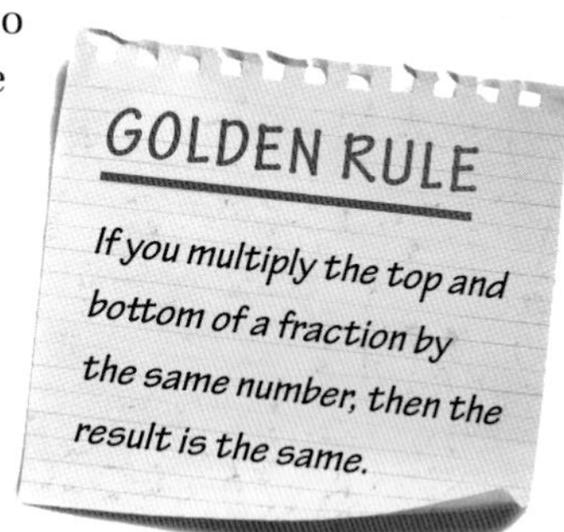

So, for example, $\frac{1}{2} = \frac{4}{8}$ is true, because the second fraction comes from multiplying the top and bottom of the first fraction by 4. A less obvious example is that $\frac{6}{7} = \frac{36}{42}$, as we can see by multiplying the top and bottom of the first fraction by 6. In terms of cakes, the golden rule tells us how to slice the helping of cake into thinner pieces, while keeping the overall helping the same size.

Simplifying fractions

Here is some jargon: the top of a fraction is known as the *numerator*; the bottom is called the *denominator*. The golden rule says that if you have a fraction then multiplying the numerator and denominator by the same number may change the superficial appearance of the fraction, but not its underlying value. Now, this rule also works backwards: if instead you *divide* the top and bottom by the same number, then the fraction's value is similarly unaffected.

So if we start with a fraction such as $\frac{15}{20}$, and then we divide both the top and bottom by 5, we see the fraction in a new form: $\frac{3}{4}$. The useful thing about this is that the numbers on top and bottom have become smaller. The fraction therefore seems simpler. This process of dividing top and bottom by the same number is known as *simplifying* the fraction.

A nice thing about $\frac{3}{4}$ is that there is no number which can divide both the top and the bottom: this version of the fraction is as simple as it gets. We might say that it has been *fully simplified*.

When writing a fraction it is usually good practice to present it in fully simplified form. This means checking whether there is any number which divides both the top and bottom, and if there is, then dividing top and bottom by it. So starting with $\frac{10}{12}$, we might notice that the top and bottom are both divisible by 2. Dividing top and bottom produces $\frac{5}{6}$, which is now in fully simplified form.

The ultimate in simplification is when we have a fraction such as $\frac{12}{4}$. Here the top and the bottom are both divisible by 4. Dividing top and bottom gives $\frac{3}{1}$. But what is $\frac{3}{1}$? Well, it is $3 \div 1$, which is 3. Whenever there is only a 1 on the bottom of a fraction, the bottom effectively disappears, leaving just the top on its own as a whole-number answer. (Going the other way, if we want to write 5 as a fraction instead of a whole number, we can just put a 1 underneath it, to get $\frac{5}{1}$. The golden rule then tells us that this is the same as $\frac{10}{2}$, $\frac{15}{3}$, and other variations on the theme. Notice that this matches our intuition, as $10 \div 2$ and $15 \div 3$ are indeed equal to 5.)

PRACTISE SIMPLIFYING FRACTIONS IN QUIZ 2.

Top-heavy fractions

Most of the fractions we have seen so far in this chapter have had a smaller number on top of larger one. But there is no law that says it must be so: $\frac{3}{2}$ is a perfectly valid fraction too, and everything that we have said so far applies equally to 'top-heavy' fractions like this (as does everything we say in the next chapter).

Most scientists don't bat an eyelid at fractions like $\frac{3}{2}$. But in the wider world they can seem a bit strange. Instead of saying 'three halves of an hour', most people would say 'one hour and a half', which might be written like this: $1\frac{1}{2}$.

Expressions like $\frac{3}{2}$ are accurately described as *top-heavy*, or, somewhat unfairly, as *improper* fractions. An expression like $1\frac{1}{2}$ is called a *mixed number*. Notice that there is an invisible '+' sign here: $1\frac{1}{2}$ is really the same thing as $1 + \frac{1}{2}$.

So much for the jargon. The point is that $1\frac{1}{2}$ and $\frac{3}{2}$ are actually the same thing. They are just presented in slightly different ways. The next challenge, then, is to be able to translate between the languages of improper fractions and mixed numbers.

Suppose we are given a top-heavy fraction such as $\frac{7}{3}$ and we want to turn this into a mixed number. The process is not too tricky: in fact it comes directly from the meaning of a fraction. Don't forget that a fraction represents division. So $\frac{7}{3}$ is exactly the same thing as $7 \div 3$. Having said that, let's calculate $7 \div 3$. The answer is that 3 goes into 7 two times, leaving a remainder of 1. The 2 is the whole number part of the answer, and the remainder 1 then goes on top of the fractional part, while 3 goes on the bottom, giving $2\frac{1}{3}$. (We saw this method in the chapter on division.) The general rule is:

> *To express a top-heavy fraction (such as* $\frac{7}{3}$*) as a mixed number, the number of times the bottom fits into the top (2) goes outside the fraction as a whole number. The remainder (1) then stays on top of the fraction, and the bottom of the fraction (3) doesn't change:* $2\frac{1}{3}$.

Going the other way is easier:

> *If we are given a mixed number (such as* $2\frac{3}{4}$*) that we want to change into an improper fraction, then we start by multiplying the whole number (2) by the bottom of the fraction* (4), *which gives us 8. Then we add on the top*

of the old fraction (3), *to get the top of the new one* (11), *and the bottom* (4) *stays the same. So the answer is* $\frac{11}{4}$.

Got it? Try quiz 3.

From decimals to fractions

We have met two ways of expressing non-whole numbers: fractions and decimals. Since both of these languages are very common, it is important to know how to translate between the two.

The simplest examples are worth memorizing: $\frac{1}{2} = 0.5$, $\frac{1}{4} = 0.25$ and $\frac{3}{4} = 0.75$.

Beyond these, some techniques are required. Suppose we are given a decimal 0.4. How can we express this as a fraction? Actually, decimals are fractions already. Why is that? Well, remember that the columns to the right of the decimal point represent tenths, hundredths, thousandths, and so on (just as the columns to the left represent units, tens, hundreds, etc.). So 0.4 really is just $\frac{4}{10}$. All that remains is to simplify it, which we can do by dividing the top and bottom by 2, to give a final answer of $\frac{2}{5}$.

Let's take another example: 0.35. We know 0.35 is three tenths plus five hundredths. Or, to put it another way, it is thirty-five hundredths: $\frac{35}{100}$. We can simplify this by dividing the top and bottom by 5, to get $\frac{7}{20}$.

In the last example, 0.35 had two decimal places, so we had to express it in hundredths initially, before simplifying. Similarly, if we want to convert 0.375 to fractional form, we need to express it first as thousandths: $\frac{375}{1000}$. The we can simplify by dividing the top and bottom by 5 three times, to get $\frac{3}{8}$.

From fractions to decimals

Have a go at this technique in quiz 4.

Suppose we are given a fraction and want to express it as a decimal. First let's take a nice friendly one, $\frac{3}{4}$, and pretend that we don't already know what the decimal equivalent is. How might we work it out?

In the last section, before we simplified the result, the fractions we arrived at looked like this $\frac{?}{10}$, $\frac{?}{100}$, $\frac{?}{1000}$, and so on. If we can convert $\frac{3}{4}$ into a fraction of one of these types, then we will nearly be there. So we need to apply this chapter's golden rule. We will try these possibilities in turn.

The bottom of our fraction is currently 4. But we want to change it to 10, or 100, or 1000, etc. Unfortunately, there is no whole number we can multiply 4 by to get 10. So we cannot express $\frac{3}{4}$ as tenths. Moving on, however, there is a number we can multiply 4 by to get 100, namely 25. So applying the golden rule, we multiply the top and bottom of $\frac{3}{4}$ by 25 to get $\frac{75}{100}$. It is now easy to recognize that this is the decimal 0.75.

Recurring decimals

Now, we run into an awkward fact. Some fractions have decimal representations which look a bit strange. Start with $\frac{1}{3}$. If you type [1] [÷] [3] [=] into your calculator, you should get an answer of 0.3333333333. In fact this is not an exact answer; the string of 3s really goes on forever. Try decimal short division for $1 \div 3$ and see what happens!

This is what is known as a *recurring decimal*, meaning that it gets stuck in a repeating pattern that goes on forever. A lot of fractions do this. For example, $\frac{1}{6} = 0.1666666666\ldots$

When calculating with these sorts of numbers, there is no choice but to round them off, after a number of decimal places, as the calculator does. However, there is a special notation for recurring decimals: a dot over the repeating number. So we would write $\frac{1}{3} = 0.\dot{3}$ and $\frac{1}{6} = 0.1\dot{6}$.

Some numbers have more complex repeating patterns. For instance $\frac{1}{7} = 0.142857142857142857\ldots$, which we would express as $\frac{1}{7} = 0.\overline{142857}$.

Every fraction will have an expression as either a terminating decimal (such as 0.51) or a recurring decimal (such as $0.5\dot{1}$). There are other numbers where the decimal expansion goes on forever without ever repeating; these are the so-called *irrational numbers*. Famous examples are π (see *Circles*) and $\sqrt{2}$ (see *Pythagoras' theorem*).

TRY EXPRESSING FRACTIONS AS DECIMALS IN QUIZ 5.

Sum up *Once you have got to know them, fractions really are a piece of cake!*

Quizzes

1 Different yet the same. Write $\frac{3}{4}$ as:

a Twelfths

b Eighths

c Sixteenths

d Twentieths

e Hundredths

2 Simplify

a $\frac{3}{6}$

b $\frac{10}{15}$

c $\frac{12}{28}$

d $\frac{25}{30}$

e $\frac{80}{100}$

3 Top-heavy fractions to mixed numbers

a $\frac{7}{4}$

b $\frac{10}{3}$

c $\frac{9}{4}$

d $\frac{9}{2}$

e $\frac{23}{7}$

4 Decimals to fractions

a 0.9

b 0.6

c 0.95

d 0.625

e 0.875

5 Fractions to decimals (possibly recurring)

a $\frac{4}{5}$

b $\frac{2}{3}$

c $\frac{3}{16}$

d $\frac{2}{7}$

e $\frac{5}{9}$

6 Which numbers on the bottom of fractions produce recurring decimals? Experiment and see!

Arithmetic with fractions

- *Knowing how to add and subtract fractions*
- *Cancelling fractions, to speed up multiplication*
- *Understanding how to divide fractions*

In the last chapter we saw how to represent numbers as fractions, noticing particularly that one number has many different fractional representations. In this chapter we will see how to do arithmetic with fractions: adding, subtracting, multiplying and dividing them.

Surprisingly, it is addition and subtraction which are trickier in this context: multiplication and division are fairly straightforward. A common error is to see $\frac{1}{2} + \frac{3}{4}$ and add the top and bottoms separately to get $\frac{4}{6}$. This is certainly wrong (think of half a cake being added to another three quarters of a cake). So let's jump in at the deep end, and get started adding fractions the correct way.

Adding fractions

Everyone can agree that adding fractions is easy, in some circumstances at least. What is one ninth plus one ninth? Two ninths, of course. What is $\frac{2}{17} + \frac{12}{17}$? It must be $\frac{14}{17}$. As long as you can add ordinary whole numbers, you can add these sorts of fractions. What makes these so simple is that these fractions all have identical bottom numbers. (In the jargon, they have a 'common denominator'.)

EASY? TRY QUIZ 1.

Changing the bottom numbers

The key to adding all fractions is to be able to transform any addition into an easy one like those we have just looked at. Take a simple example: what is $\frac{1}{2} + \frac{1}{4}$? Slicing up a cake quickly reveals the answer: $\frac{3}{4}$. But we want a calorie-controlled way to see this directly from the numbers, skipping the cakes.

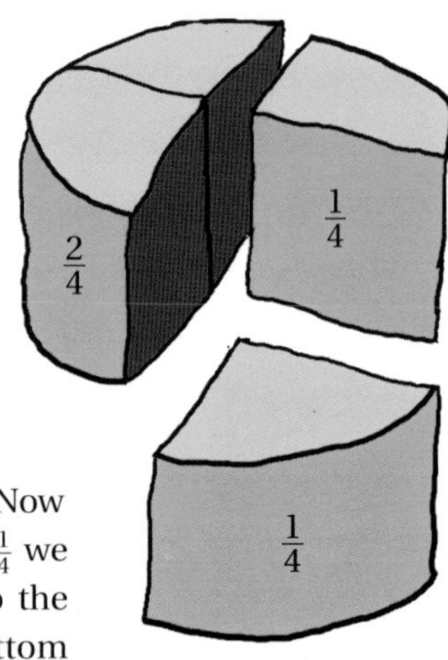

The crucial observation is that the half can be broken down into two quarters, using the previous chapter's golden rule for fractions. Now we can rewrite the question. Instead of $\frac{1}{2} + \frac{1}{4}$ we can write it as $\frac{2}{4} + \frac{1}{4}$. This brings us back to the familiar ground of fractions with the same bottom number, which we can then add straightforwardly. From this we get the golden rule for this chapter.

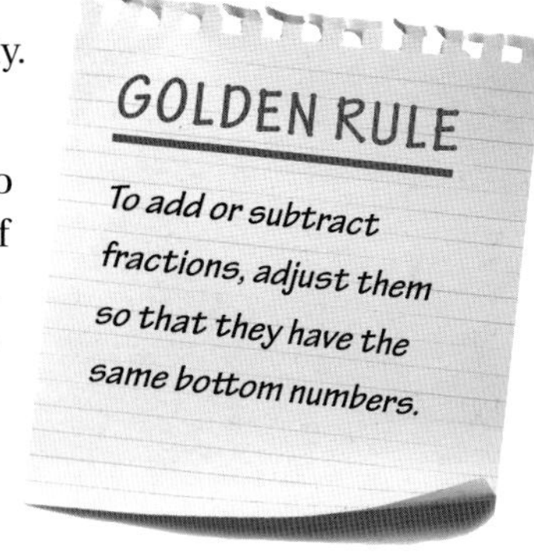

In the last example, we were able to split a half into two quarters. A similar trick will work with lots of other fractions. Suppose we want to calculate $\frac{5}{12} + \frac{1}{6}$. This time we can break the single sixth into two twelfths (using the previous chapter's golden rule), and so change the question to $\frac{5}{12} + \frac{2}{12}$. Once again we are back in the comfortable realm of matching bottoms! Another example is $\frac{1}{2} + \frac{3}{8}$. This time, we split the half into four eighths: $\frac{4}{8} + \frac{3}{8}$.

GET THIS TRICK? THEN TRY IT OUT IN QUIZ 2.

When cakes collide

A thornier type of problem is one like this: $\frac{1}{2} + \frac{1}{3}$. The trouble with this is that we cannot split a half into thirds, or a third into halves. Neither fits into the other. This time, to get to a sum in which the two fractions have matching bottom numbers we are going to have to alter both fractions. But which numbers should we multiply their tops and bottoms by? Let's try some out. If we start with $\frac{1}{2}$, applying the previous chapter's golden rule with 1, 2, 3, 4, 5 in turn produces the fractions $\frac{1}{2}$, $\frac{2}{4}$, $\frac{3}{6}$, $\frac{4}{8}$, $\frac{5}{10}$, etc., all equal to $\frac{1}{2}$. Meanwhile, starting with $\frac{1}{3}$ produces $\frac{1}{3}$, $\frac{2}{6}$, $\frac{3}{9}$, $\frac{4}{12}$, $\frac{5}{15}$, etc. Looking at these two sequences, we might spot that they both contain a fraction with 6 on the bottom. This is promising, as it allows us to rewrite $\frac{1}{2} + \frac{1}{3}$ as $\frac{2}{6} + \frac{3}{6}$, and then proceed as before.

Why did the number 6 work so well here? Looking at the bottom row of the sequence $\frac{1}{2}$, $\frac{2}{4}$, $\frac{3}{6}$, $\frac{4}{8}$, $\frac{5}{10}$, what we see is nothing other than the two times table. Similarly the bottom row of $\frac{1}{3}$, $\frac{2}{6}$, $\frac{3}{9}$, $\frac{4}{12}$, $\frac{5}{15}$ is the three times table. The number 6 worked perfectly then because it features in *both* times tables. (In fact it is the first number to feature in both; this is the so-called 'lowest common multiple'.)

So, when faced with the sum of fractions, $\frac{2}{3} + \frac{1}{5}$, the first thing to do is to find a number which features in both the three and five times tables. Thinking about it, 15 is such a number. Then we can apply the golden rule twice, to turn $\frac{2}{3} + \frac{1}{5}$ into a sum of two fractions with 15 on their bottoms. To get 15 on the bottom of $\frac{2}{3}$, we must multiply by 5. So, according to the golden rule, we must multiply the top by 5 too: $\frac{10}{15}$. Similarly, to get 15 on the bottom of $\frac{1}{5}$, we have to multiply top and bottom by 3. This produces $\frac{3}{15}$, and our sum becomes $\frac{10}{15} + \frac{3}{15}$, which we can easily finish off to get $\frac{13}{15}$.

The final step, when adding fractions, is to make sure that the answer is fully simplified, since this doesn't happen automatically. For instance if we add $\frac{1}{6} + \frac{1}{3}$, by following the rules above, we can turn this into $\frac{1}{6} + \frac{2}{6}$, which then becomes $\frac{3}{6}$. But $\frac{3}{6}$ can be simplified to give $\frac{1}{2}$, which is the final answer.

Subtracting fractions

The rules for subtracting fractions are almost identical to those for adding them. The only difference is that, where previously we added, now we subtract. Genius! So, to take an example, if we have $\frac{3}{4} - \frac{1}{8}$, then exactly as before we change this to $\frac{6}{8} - \frac{1}{8}$. We can then evaluate it as $\frac{5}{8}$. Since this can't be simplified, we have finished.

PRACTISE ADDITION AND SUBTRACTION IN QUIZ 3.

If you are comfortable with adding fractions, then subtraction should be a piece of cake.

Multiplying fractions

While adding and subtracting fractions is a slightly tricky procedure, multiplying them is straightforward. For instance, to calculate $\frac{2}{3} \times \frac{1}{5}$ all we do is multiply the numbers on the top (2×1), and the numbers on the bottom (3×5), and then put the answer back together: $\frac{2}{3} \times \frac{1}{5} = \frac{2 \times 1}{3 \times 5} = \frac{2}{15}$.

The procedure may be easy, but what is actually going on, in the language of slices of cake? Well, suppose you have sliced a cake into fifths. Calculating $2 \times \frac{1}{5}$ would correspond to a serving of two such slices, amounting to $\frac{2}{5}$ of a cake. Similarly, the calculation $\frac{2}{3} \times \frac{1}{5}$ gives the size of a serving which comprises $\frac{2}{3}$ of one those slices. As we have seen, this amounts to $\frac{2}{15}$ of the whole cake. This is a case of multiplication being implied by one short English word: 'of'. What $\frac{2}{3} \times \frac{1}{5}$ really means is $\frac{2}{3}$ *of* $\frac{1}{5}$.

Cancelling

Of course, after multiplying fractions you may need to simplify the answer (as you also need to do when adding or subtracting them). In this case, though, a shortcut is sometimes possible.

To see how, let's take another example: $\frac{2}{7} \times \frac{7}{9}$. The rules for multiplying fractions (together with knowledge of times tables) make it simple to come up with an answer: $\frac{14}{63}$. The final step is to simplify this fraction. So the question is: is there any number which can divide both the top number and the bottom number? A little reflection will reveal an answer: 7. So dividing top and bottom by 7 produces the final result: $\frac{2}{9}$.

That all worked perfectly well, but going back to the original question $\left(\frac{2}{7} \times \frac{7}{9}\right)$ we can actually see the final answer $\left(\frac{2}{9}\right)$ immediately if we know how to look for it, without having to work through all the multiplication and simplifying.

The question amounts to this: $\frac{2\times7}{7\times9}$. On the bottom of this fraction, even if we have forgotten all our times tables, we can see straight away that this number is divisible by 7. And the same goes for the top. So we can move directly to simplifying, by dividing out by 7 immediately, which we might write like this: $\frac{2\times\cancel{7}}{\cancel{7}\times9} = \frac{2}{9}$.

This process is often known as *cancelling*, and it is a great labour-saving device. It comes into its own when multiplying more than two fractions together. Suppose we are faced with $\frac{2}{5} \times \frac{3}{7} \times \frac{7}{8} \times \frac{5}{3}$. We can immediately set to cancelling the 3 on the top with that on the bottom, the 7 on top with the one below, and similarly for the 5s: $\frac{2\times\cancel{3}\times\cancel{7}\times\cancel{5}}{\cancel{5}\times\cancel{7}\times8\times\cancel{3}} = \frac{2}{8}$. This can finally be simplified to $\frac{1}{4}$. Once you get the gist of this, it is much easier than evaluating it as $\frac{210}{840}$, and then fiddling around trying to simplify it from there.

As another example, take $\frac{7}{15} \times \frac{5}{28}$. This time, the numbers on the top and bottom are all different. Yet some cancelling is possible nevertheless. The 15 on the bottom is divisible by 5, which occurs on the top. So we can cancel those out, leaving 3 on the bottom (because $3 \times 5 = 15$). Similarly, we can cancel the 7 on the top with the 28 on the bottom, to leave 4 on the bottom $\frac{\cancel{7} \times \cancel{5}}{\cancel{15}\,3 \times \cancel{28}\,4}$.

NOW HAVE A GO AT QUIZ 4.

When everything is cancelled from one line, as happens on the top in this case, there is always an invisible 1 which remains. So the final answer in this case is $\frac{1}{12}$.

Dividing fractions

What does it mean to ask what $10 \div 5$ is? I'm not interested in the answer (that's easy), but rather the meaning of the question. Well, the answer should be the number which tells us how many times 5 fits inside 10. To put it another way, the question we have to solve is $5 \times \boxed{?} = 10$.

So far, so good. But what happens when we ask the same question of fractions: what is $\frac{1}{2} \div \frac{1}{6}$? According to the same philosophy, it ought to be the number of times that $\frac{1}{6}$ fits inside $\frac{1}{2}$. To put it another way, $\frac{1}{6} \times \boxed{?} = \frac{1}{2}$. The answer to this is 3.

This is all perfectly correct, but it is a common source of confusion that dividing one small number by another small number can produce something much *bigger*. Shouldn't division always produce smaller results? The answer is a categorical 'no'! Dividing by a large number gives a smaller answer, but dividing by small numbers (meaning smaller than 1) gives *bigger* answers.

To go back to $10 \div 5$, it is no coincidence that the answer here is the same as the answer to $10 \times \frac{1}{5}$. Why? Because $10 \div 5$ means the same as $\frac{10}{5}$, as does $10 \times \frac{1}{5}$.

What this suggests is the following rule for dividing fractions:

> *To divide one fraction by another, turn the second one upside down, and then multiply.*

So we would write the above example out as $\frac{1}{2} \div \frac{1}{6} = \frac{1}{2} \times \frac{6}{1} = \frac{3}{1} = 3$.

Another example might be $\frac{3}{7} \div \frac{3}{4}$. According to the rule for fractional division, this is equal to $\frac{3}{7} \times \frac{4}{3}$, which, after cancelling, becomes $\frac{4}{7}$.

Now have a go at dividing fractions in quiz 5.

A word of warning: sometimes fractional division can look very different, when the division symbol is itself replaced with a fraction. For instance, the last example might also be written as $\frac{\frac{3}{7}}{\frac{3}{4}}$. Don't be put off by this – the meaning is exactly the same as before. And what if we saw something like $\frac{\frac{3}{7}}{2}$? Well this has the same meaning as $\frac{3}{7} \div 2$, which is the same as $\frac{3}{7} \div \frac{2}{1}$, and so equal to $\frac{3}{7} \times \frac{1}{2}$.

Sum up *Fractional arithmetic is a piece of cake, so long as you remember the rules!*

Quizzes

1 When bottoms match

a $\frac{1}{15} + \frac{2}{15}$

b $\frac{7}{100} + \frac{22}{100}$

c $\frac{13}{29} + \frac{12}{29}$

d $\frac{10}{33} + \frac{17}{33}$

e $\frac{10}{999} + \frac{18}{999}$

2 Mismatching bottoms

a $\frac{1}{6} + \frac{1}{18}$

b $\frac{1}{3} + \frac{7}{12}$

c $\frac{2}{5} + \frac{3}{10}$

d $\frac{7}{10} + \frac{3}{100}$

e $\frac{3}{5} + \frac{9}{25}$

3 Double trouble

a $\frac{2}{5} - \frac{1}{3}$

b $\frac{1}{3} + \frac{1}{4}$

c $\frac{3}{4} - \frac{1}{5}$

d $\frac{1}{6} + \frac{3}{4}$

e $\frac{4}{5} - \frac{1}{6}$

4 Cancel down first!

a $\frac{1}{3} \times \frac{3}{4}$

b $\frac{3}{5} \times \frac{2}{3}$

c $\frac{3}{4} \times \frac{2}{5} \times \frac{5}{3}$

d $\frac{5}{8} \times \frac{4}{15}$

e $\frac{7}{15} \times \frac{2}{21} \times \frac{3}{8}$

5 Fractions turn upside-down

a $\frac{1}{4} \div \frac{1}{3}$

b $\frac{2}{5} \div \frac{4}{5}$

c $\frac{\frac{3}{8}}{\frac{6}{7}}$

d $\frac{\frac{5}{6}}{\frac{10}{21}}$

e $\frac{7}{15} \times \frac{1}{14} \div \frac{3}{5}$

Powers

- *Understanding what powers are and what they are useful for*
- *Understanding how bank interest works*
- *Understanding exponential growth*

Ultimately, multiplication is repeated addition. So 5×4 is four 5s added together: $5 + 5 + 5 + 5$. In the same way, we can step up a level and look at powers – the name for repeated multiplication. So 5^4 (that's pronounced '5 to the power 4') is four 5s multiplied together: $5^4 = 5 \times 5 \times 5 \times 5$.

What is the point of these powers? One answer to that question comes from geometry, where they crop up all the time, as we shall see in the chapter on area and volume. In fact, there are special names for certain powers, which already suggest geometrical ideas.

Raising to the power 2 is known as 'squaring', so 5^2 or 5×5 is also known as 'five squared'. The reason for this is that if you draw a square whose length is 5cm, then its area is 5^2cm^2.

In the same way, raising to the power 3 is known as 'cubing'. So 5^3 (or $5 \times 5 \times 5$) is 'five cubed'. Why? Because a cube with sides each 5cm long has volume 5^3cm^3 (see *Area and volume*).

NOW HAVE A GO AT QUIZ 1.

Calculators and powers

Powers are usually written as superscripts: 5^{12}. If you are working on a computer, however, the symbol ^ is often used instead: 5^12. As we shall see shortly, powers grow very large very quickly, so using a calculator or computer is often unavoidable. Most calculators will have a button marked [x^y] (or perhaps [$x^\square$]) with which to calculate powers. (If this symbol appears above another button, you may need to press the 'SHIFT' or 'Second Function' button first to access this function.) To calculate 5^{12}, you would need to press [5] [x^y] [1] [2] [=]

TRY THIS OUT IN QUIZ 2.

Sessa's chessboard

Another word for a power is 'exponent'. So in 5^3 the *exponent* is 3.

Exponential growth is famous for being extremely fast. In fact, there is a myth about this, which concerns an Indian wise man called Sessa, said to be the original inventor of chess. When he took his invention to show the King, His Majesty fell in love with the game. In fact, he was so impressed that he declared that he would give Sessa anything he wanted as a reward, anything at all. Sessa's reply was the stuff of legend.

Laying out his chessboard in front of him, he said that he would like 1 grain of wheat on the first square of the board. On the second square, he would like 2 grains of wheat, and on the next square 4 grains, and then 8, and so on, continuing in the same manner until each square was full.

The King was annoyed that his generosity should be ridiculed, thinking that a few grains of wheat was not much to request of a great leader. But how many grains of wheat would the board require?

TRY OUT QUIZ 3!

In quiz 3 you have to fill in the numbers of Sessa's chessboard as far as you can, without a calculator. If you can get as far as the 21st square, you are doing well! A chessboard has 64 squares: if you can complete them all, you are doing better than some pocket calculators.

Exponential growth

In the story of Sessa's chessboard, there is 1 grain of wheat on the first square. It is a mathematical convention that any number raised to the power of 0 gives 1. So we can write this as 2^0. Don't worry if this seems a little strange; it is just a rule we adopt to make sense of otherwise meaningless expressions like 2^0. It turns out to be convenient, as we don't have to keep talking about 0 separately, as a special case.

On the second square of the board, there were 2 grains, that is to say 2^1. On the next, there were 4, that is 2^2, and on the next $8 = 2^3$, and so on. It is the power (or exponent) of 2 which grows by 1 from square to square, making this a classic example of 'exponential growth'. As the story shows, exponential growth can be very fast indeed. The King certainly had to give more than he bargained for: by the 51st square the King would have to hand over the entire global wheat harvest for 2007.

Exponential decay

We have seen that powers can grow very quickly. But it is not always so. It is certainly true that if we take a number such as 2, and multiply by itself, then the answer is bigger. As Sessa knew, the more times we multiply it by itself, the bigger it gets. But not all numbers are like this. If we take a number like $\frac{1}{2}$, then when we square it we get something smaller: $\left(\frac{1}{2}\right)^2 = \frac{1}{2} \times \frac{1}{2} = \frac{1}{4}$. If we cube it, we get something smaller still: $\left(\frac{1}{2}\right)^3 = \frac{1}{8}$, and so on.

This is an example of *exponential decay*: the flipside of exponential growth. As with exponential growth, the process is very quick indeed: $\left(\frac{1}{2}\right)^{20}$ is less than a millionth.

Where is the boundary between exponential growth and exponential decay? Which numbers grow big when repeatedly multiplied by themselves, and which shrink away towards 0? The dividing line is the number 1. When you raise 1 to a power, nothing happens as $1 \times 1 \times 1 \times \cdots \times 1$ is always 1, no matter how many multiplications there are. But anything less than 1 (but still greater than 0) will decay, and anything bigger than 1 will grow.

What is surprising is how quickly this happens, even for numbers which are only just on one side or the other of 1. For example, 0.9^{22} is already less than 0.1, while 1.1^{22} is more than 8.

NOW HAVE A GO AT QUIZ 4.

The arithmetic of powers

For the final part of this chapter, let's look a little deeper at the arithmetic of powers. How do we work out $2^4 \times 3^4$? Writing this out, we get:

$$\underbrace{2 \times 2 \times 2 \times 2} \times \underbrace{3 \times 3 \times 3 \times 3}$$

So that's four 2s multiplied by four 3s. But we know that when we multiply things together the order doesn't matter. So we can rearrange this as:

$$\underbrace{2 \times 3} \times \underbrace{2 \times 3} \times \underbrace{2 \times 3} \times \underbrace{2 \times 3}$$

This is the same as $6 \times 6 \times 6 \times 6$, or 6^4.

This is an example of a general rule that, for any numbers *a*, *b*, *c*,

$$a^c \times b^c = (a \times b)^c$$

What happens when we multiply two powers of the same number together? For example, what is $2^3 \times 2^4$? Writing out the powers as multiplication, the answer is, $\underbrace{2 \times 2 \times 2} \times \underbrace{2 \times 2 \times 2 \times 2}$. Counting up the number of 2s gives the answer: 2^7.

It is no coincidence that $2^3 \times 2^4 = 2^7$ and also $3 + 4 = 7$. This is one of the rules (or laws) for powers. It says that, for any numbers a, b, c, it will always be true that:

$$a^b \times a^c = a^{b+c} \quad \text{(The first law of powers)}$$

What this says is that to multiply two powers together you *add* their exponents (but this only works if they are powers of the same number!).

Another rule or law tells us what happens when we look at a power of a power, for example: $(2^3)^4$. What is this? Writing it out, we get, $2^3 \times 2^3 \times 2^3 \times 2^3$, and expanding these we get:

$$\underbrace{2 \times 2 \times 2} \times \underbrace{2 \times 2 \times 2} \times \underbrace{2 \times 2 \times 2} \times \underbrace{2 \times 2 \times 2}$$

There are twelve 2s here altogether, because $3 \times 4 = 12$, so we get an answer of 2^{12}. This gives us the general rule, for any numbers a, b, c,

$$(a^b)^c = a^{b \times c} \quad \text{(The second law of powers)}$$

TRY USING THE LAWS OF POWERS IN QUIZ 5.

Sum up *From exponential growth, to the laws of powers, powers are subtle calculations. But remember, really there's nothing more to them than repeated multiplication!*

Quizzes

1 The power of powers: write these out in full.

a 3^3
b 6^2
c 5^3
d 3^4
e 6^3

2 Even more powerful powers: calculate the values.

a 5^5
b 6^6
c 12^4
d 125^2
e 89^5

3 Continue the sequence of numbers from Sessa's chessboard.
1, 2, 4, 8, 16, . . .

4 Work these out leaving the answer as a fraction.

a $\left(\frac{1}{10}\right)^4$
b $\left(\frac{1}{2}\right)^7$
c $\left(\frac{2}{3}\right)^4$
d $\left(\frac{1}{100}\right)^3$
e $\left(\frac{3}{4}\right)^4$

5 Use the arithmetic of powers to express each of these as a single number to a single power (e.g. 8^{16}).

a $2^{14} \times 2^8$
b $(2^5)^3$
c $3^5 \times 5^5$
d $13^{19} \times 13^7$
e $25^{19} \times 4^{19}$

The power of 10

- *Knowing the names of extremely large numbers*
- *Being able to write very large and very small numbers*
- *Understanding the metric system of measurements*

There is something special about the number 10. The numbers 0–9 each have their own individual symbol. But when we reach 10 its symbol is made up from those for 0 and 1. This simple observation cuts to the very heart of the modern way of representing numbers. It was not always like this, as anyone familiar with Roman numerals knows.

It is not just 10 which is significant, but all the *powers* of 10. These are 100, 1000, 10,000, 100,000, and so on. These are special in our way of writing numbers, since they mark the points where numbers become longer: while 99 is two digits long, 100 is three, while 999 is three digits long, 1000 is four, and so on.

The number 1000 is $10 \times 10 \times 10$, which, in the language of powers is 10^3. (See the previous chapter for a general discussion of powers.) The important observation is this: the power of 10, in this case 3, actually counts the zeros. So 10^3 is the same as 1 followed by three zeros. This becomes very useful as the numbers get larger. The expression 10^{10} can be read and digested much more easily than if you were left to count the zeros yourself: 10,000,000,000.

Powers of 10 are also the points where new names for numbers appear. If we scroll down the powers of 10, the first few are simple enough:

10^0	1	one
10^1	10	ten
10^2	100	hundred
10^3	1000	thousand

After a thousand, new names appear every three steps, with ten and a hundred filling the intermediate gaps. So:

10^4	ten thousand
10^5	hundred thousand
10^6	million

10^7	ten million
10^8	hundred million
10^9	billion

It's the multiples of three which are important from the point of view of naming numbers.

10^3	thousand
10^6	million
10^9	billion
10^{12}	trillion
10^{15}	quadrillion
10^{18}	quintillion
10^{21}	sextillion
10^{24}	septillion

A word of caution here: in the past there was some disagreement across the Atlantic about what constituted a 'billion'. Americans have always considered a billion to be a thousand million (that is, 10^9) while the British used to use the same word to mean a million million (that is, 10^{12}). That is no longer the case. Today, the system above is universal in the English-speaking world. However, the disparity is worth remembering, if you are ever reading British documents dating from before 1974. In other languages, systems vary. In French, for example, 10^6 is *un million*, 10^9 is *un milliard*, and 10^{12} is *un billion*. In Japanese, Chinese and Korean, the basic unit is 10^4 rather than 10^3 with new numerical names appearing at 10^4, 10^8, 10^{12}, 10^{16}, and so on. Translators beware!

COUNT THOSE ZEROS! TRY QUIZ 1.

The metric system

Often when we see numbers written down, there are a few letters after them: for example, 5kg, 10s, 12cm. These letters are different from the ones that appear in algebra (see *Algebra*). Instead these are *units*, and their purpose is to define exactly what the numbers are measuring, whether that be mass, time, distance or something else, and the scale being used to measure it.

There is a whole army of units that people use to measure everything from humidity (g/m^3, that is, *grams per cubic metre*) to the spiciness of chillies (SHU, that is, *Scoville heat units*). It would not be practical or useful to attempt a complete list!

Nevertheless, there is something important to say about the way that a certain class of unit relates to the powers of 10 we have just been looking at. This is known as the *metric system*, and it is based on *metres* for distance (rather than inches or miles), *grams* for weight (rather than ounces or stones), *seconds* for time (rather than minutes or years). Other units are then built out of these. A *litre*, for example, is $100cm^3$ (see *Area and volume*), while the standard measure of force is the *Newton*, defined to be $1kg\ m/s^2$.

Let's take an example, to see how the system works. A *gram* is a unit of weight. One gram on its own is not very much. So if we want to measure people, cars or planets, a gram doesn't seem a very satisfactory starting point. However, there are prefixes which can be put in front of the word 'gram', to make the unit bigger. One is 'kilo' which means a thousand. So 1 kilogram is the same thing as 1000 grams. The kilogram is a sensible unit for measuring the mass of a person, for example. Just as with the names of numbers, new metric prefixes generally occur every multiple of 3:

10	deca-
10^2	hecto-
10^3	kilo-
10^6	mega-
10^9	giga-
10^{12}	tera-
10^{15}	peta-
10^{18}	exa-
10^{21}	zetta-
10^{24}	yotta-

Many modern scientists might frown at the first two in this list, but mechanical engineers do sometimes discuss force in terms of *decanewtons* (daN), while meteorologists occasionally measure atmospheric pressure in *hectopascals* (hPa). But it is for larger numbers that the system really becomes useful. It is very common to measure distances in *kilometres* (km), and the resolution of a digital camera might be 5 *megapixels* (MP), meaning that it contains 5 million individual image sensors (pixels).

Unless you are an astrophysicist measuring the weight of stars, the largest of these prefixes you will probably ever need is 'tera-', or 10^{12}. It is quite common for computers to have 1 *terabyte* (TB) disk drives now.

If you wanted to measure the weight of a car, the *megagram* would be a sensible unit to use. It just happens that the megagram more commonly goes by the name of *tonne*, meaning a million grams, or equivalently, a thousand kilograms.

IF YOU KNOW YOUR PREFIXES, HAVE A GO AT QUIZ 2!

Small things

Everything we have said for the very large goes equally well for the very small. We can use *negative powers* of 10 to represent small numbers. To start with, 10^{-1} means $\frac{1}{10}$, or equivalently 0.1. Similarly, 10^{-2} is $\frac{1}{10^2}$. which is $\frac{1}{100}$, or 0.01, and 10^{-3} is $\frac{1}{10^3}$ which is $\frac{1}{1000}$, or 0.001, and so on.

A quick rule, as before, is that the negative power counts the number of zeros, with the important caveat that a single zero before the decimal point must be included in the count. With this said, it is easy to see that 10^{-6} is one *millionth*, 10^{-9} is one *billionth*, and so on. Writing these in decimals, we get $10^{-6} = 0.000001$ and $10^{-9} = 0.000000001$.

There are also metric prefixes for the small numbers:

10^{-1}	deci-
10^{-2}	centi-
10^{-3}	milli-
10^{-6}	micro-
10^{-9}	nano-

NOW HAVE A GO AT QUIZ 3.

10^{-12}	pico-
10^{-15}	femto-
10^{-18}	atto-
10^{-21}	zepto-
10^{-24}	yocto-

Again the first of these is less commonly used than the second and third, although the standard measure of loudness, the *decibel,* was originally defined as one tenth of a *bel* (a unit which has long since fallen out of favour). We commonly use a *centimetre* (cm), which is a hundredth of a metre, and a *millilitre* (ml), which is one thousandth of a litre. A pill might contain 5 *micrograms* (μg) of vitamin D, and we have all heard the hype surrounding *nanotechnology,* meaning engineering which takes place on the scale of *nanometres* (nm). Getting very small, 3 *picoseconds* (ps) is the time it takes a beam of light to travel 1 millimetre.

Standard form

Metric prefixes and funny names like 'sextillion' are sometimes useful, and are good fun. But actually, with powers of 10 at our disposal, they are not strictly necessary.

Your calculator, for instance, can function perfectly well without them. Type in two large numbers to be multiplied together, perhaps 20,000,000 × 80,000,000. How does your calculator display the result? Mine displays '1.6×10^{15}'. We could translate this as 1,600,000,000,000,000 or '1.6 quadrillion'.

But actually the expression '1.6×10^{15}' is shorter and easier to understand than either of these. My calculator is taking advantage of powers of 10, to express this very large number in an efficient and compact way: 1.6×10^{15}. This way of representing numbers is known as *standard form.*

It is an essential skill to be able to move back and forth between standard form and traditional decimal expressions, and that is the final topic we shall explore in this chapter.

In technical terms, a number is in standard form if it looks like this: $A \times 10^B$, where A is a number between 1 and 10 (not necessarily a whole number), and B is a positive or negative whole number. An example is 3.13×10^4. Let's translate this back into ordinary notation. Above, we saw that the power of 10 can correspond to the number of zeros on the end (so $10^3 = 1000$ for instance). This is fine when we are considering 1 followed by a line of zeros, but 3.13 is not like this. We now need a slightly more sophisticated perspective. The answer comes from the chapter on multiplication, where we saw that multiplying by 10 corresponds to shifting the digits one step to the left with respect to the decimal point. So multiplying by 10^4 is equivalent to shifting to the left four times: $3.13 \rightarrow 31.3 \rightarrow 313.0 \rightarrow 3130.0 \rightarrow 31300.0$

That final number, 31,300, is our answer. (This should not come as a surprise, since $10^4 = 10{,}000$ and 31,300 is 3.13 lots of 10,000.) Another way to think of this is that the 10^4 tells us the *length* of the number. Just as 10^4 is 1 followed by four zeros, so 3.13×10^4 will be 3 followed by four other digits (of which the first two are 13).

When the power of 10 is negative, as happens in 2.83×10^{-4}, we have to shift the digits right instead of left: $2.83 \rightarrow 0.283 \rightarrow 0.0283 \rightarrow 0.00283 \rightarrow 0.000283$

In this case, it is actually easier to jump straight to the final answer, since the negative power (4 in this example) simply counts the zeros to be stuck on the front, including one zero before the decimal point as usual.

TRANSLATE NUMBERS FROM STANDARD FORM IN QUIZ 4.

From decimals to standard form

We have seen how to translate standard form into ordinary decimal notation. Now let's go in the other direction. Suppose I want to express 2000 in standard form. That's easy enough: since 2000 consists of two lots of 1000, or 10^3, it is equal to 2×10^3. This is now in standard form.

Let's take another example: 57,800. The definition of standard form dictates that the answer must look like $5.78 \times 10^?$. The only question is: what will the power of 10 be? Well, how many times would we need to shift the digits?

$$5.78 \rightarrow 57.8 \rightarrow 578 \rightarrow 5780 \rightarrow 57{,}800$$

There are four rightwards shifts there, so the answer must be 5.78×10^4. Alternatively, we could just notice that the original number is a 5 followed by 4 other digits.

The principle is the same for small numbers, such as 0.0000997. Again the standard form representation will be $9.97 \times 10^{-?}$. We just need to know the negative power of 10. It turns out that five rightwards shifts are needed:

$$9.97 \rightarrow 0.997 \rightarrow 0.0997 \rightarrow 0.00997 \rightarrow 0.000997 \rightarrow 0.0000997$$

Alternatively, we could observe that there are five zeros at the beginning of the number (including the one before the decimal point). So the answer is 9.97×10^{-5}.

NOW HAVE A GO YOURSELF IN QUIZ 5.

Using standard form for measurements

Using standard form, we can measure any distances in metres. For instance, the distance to the Sun's nearest neighbour, Proxima Centauri, is around 4×10^{16} metres. We could write this as 40 *petametres*, but this might raise a few eyebrows, although it is perfectly correct. (It would be more usual to say 4.2 *light years*, with one light year coming in at just under 10 petametres.) But actually, '4×10^{16} metres' is already a perfectly good description.

Similarly we can measure geological timescales in seconds, if we like: the Jurassic era began around 6.3×10^{15} seconds ago.

Sum up *Powers of 10 are extremely convenient for writing down very large or very small numbers!*

Quizzes

1 Write these numbers out in full.

a 7 million
b 8 billion
c 9 trillion
d 10 quadrillion
e 11 quintillion

2 Express these quantities using suitable prefixes.

a A distance of 18,000 metres
b A computer screen containing 37,000,000 pixels
c A blood cell which weighs 0.000000003 grams
d A steam-hammer which exerts a force of 900,000 newtons
e A music player whose memory is 8,000,000,000 bytes

3 Convert these numbers to words (such as a *millionth*) and also decimals (such as 0.000001).

a 10^{-3}
b 10^{-2}
c 10^{-5}
d 10^{-7}
e 10^{-12}

4 Convert these standard form numbers to ordinary decimal numbers.

a 6×10^{5}
b 2.1×10^{4}
c 8.79×10^{-6}
d 1.332×10^{-3}
e 6.71×10^{10}

5 Write these numbers in standard form.

a 800,000
b 56,000
c 0.00062
d 987,000,000
e 0.00000000111

Roots and logs

- *Understanding what roots are and what they are useful for*
- *Getting to grips with what logarithms actually mean*
- *Knowing how to switch between the language of powers, roots and logs*

A number, when squared, produces 9. What is that number? We could write this question as $?^2 = 9$. So long as we remember what* squaring *means (see* Powers *if you don't!), the answer should be obvious: 3 squared is 9 ($3^2 = 3 \times 3 = 9$), so the answer is 3. In this chapter, we will be interested in this process of squaring* backwards. *The technical term for this is* square-rooting. *So we say that 3 is the 'square root' of 9.

Square roots

There are various ways to write a square root: the commonest is to use the symbol $\sqrt{\ }$. (Don't get this confused with the symbols sometimes used to write out long division.) So we would write $\sqrt{9} = 3$.

The easiest way to figure out a square root is to flip the question round, to talk about squaring instead. So if we're asked to calculate $\sqrt{49}$, the answer is going to be the number which when squared produces 49. So we need to solve $?^2 = 49$.

If that seems clear, then have a go at Quiz 1.

This question is easy enough (I hope!). But in most cases, the square root of a whole number will not itself be a whole number. In such cases, you will need to use a calculator to get at the answer, and there is a dedicated button [√] to do the job. So, to calculate $\sqrt{15}$, you would need to type [√] [1] [5] [=], to arrive at an answer of 3.87 (to two decimal places).

Roots, roots, roots!

It is not only squaring which has a corresponding root. We can ask exactly the same thing for other powers. For example, if a box is cube-shaped and has a capacity of 64cm^3, how wide is it? Well, the volume of the box must be the width cubed (that is, multiplied by itself three times). So what we want is the number which when cubed gives 64, that is, $?^3 = 64$. This amounts to finding the *cube root* of 64. The answer is 4 because $4^3 = 4 \times 4 \times 4 = 64$. We write this as $\sqrt[3]{64} = 4$, introducing a little 3 to the root symbol.

Then the same thing works for all higher powers too. We could ask for the fourth root of 81, or $\sqrt[4]{81}$, meaning the number which when it is multiplied by itself four times gives 81.

According to this rule, where a cube root is written as $\sqrt[3]{\ }$, and a fourth root as $\sqrt[4]{\ }$, the square root symbol $\sqrt{\ }$ could equally well be written as $\sqrt[2]{\ }$. But, because it is the commonest root, it is usual practice to leave out the little '2'.

Most roots of most numbers do not produce a whole number as the answer. So, often, the safest recourse is to use the calculator. But beware: the square root button $\boxed{\sqrt{\ }}$ does *not* do the job for higher roots! There is another button for calculating higher roots, which might be indicated by $\boxed{\sqrt[x]{y}}$ or $\boxed{\sqrt[\square]{\square}}$. You might also need to press the $\boxed{\text{SHIFT}}$ or $\boxed{2^{\text{nd}}\ \text{F}^{\text{n}}}$ key to access this.

So, to calculate $\sqrt[5]{9}$, for, example, you would need to press $\boxed{5}\ \boxed{\sqrt[x]{y}}\ \boxed{9}\ \boxed{=}$ to arrive at an answer of 1.55, to two decimal places.

NOW HAVE A GO AT QUIZ 2.

Fractional powers – another way to write roots

There is another way to write roots, which does not use the root symbol ($\sqrt{\ }$), and which is worth being aware of. We can write roots as *fractional powers*. Instead of writing $\sqrt{4}$ we would write $4^{\frac{1}{2}}$, and instead of $\sqrt[3]{8}$, we would write $8^{\frac{1}{3}}$. Each time, the little number in the root symbol is written underneath a 1 (as a fraction) and then becomes a power.

The advantage to this is that it allows roots and powers to be combined quite easily. For instance, you might want first to take the cube root of 8, and then square the result. This looks cumbersome using the root notation: $(\sqrt[3]{8})^2$. In the power notation this can be written much more neatly as $8^{\frac{2}{3}}$. (This is based on the second law of powers: $(a^b)^c = a^{b \times c}$: see *Powers*.)

When faced with something like $16^{\frac{3}{2}}$. there are two things happening to the number 16. The little 2, at the bottom of the fraction, indicates not squaring but square rooting. Meanwhile the little 3 (on top of the fraction) indicates cubing (raising to the power of 3). To calculate the answer, we perform both of these steps. First calculate the root $16^{\frac{1}{2}} = 4$. Next raise that number to the power of 3, $4^3 = 64$. (In fact, the order doesn't matter, you could equally well calculate the power first: $16^3 = 4096$, and then take the root, $4096^{\frac{1}{2}} = 64$. The fact that the two answers match, and mesh so well with the second law of powers, is what makes this notation very satisfying!)

TRY SOME FRACTIONAL POWERS IN QUIZ 3.

Lovely logarithms

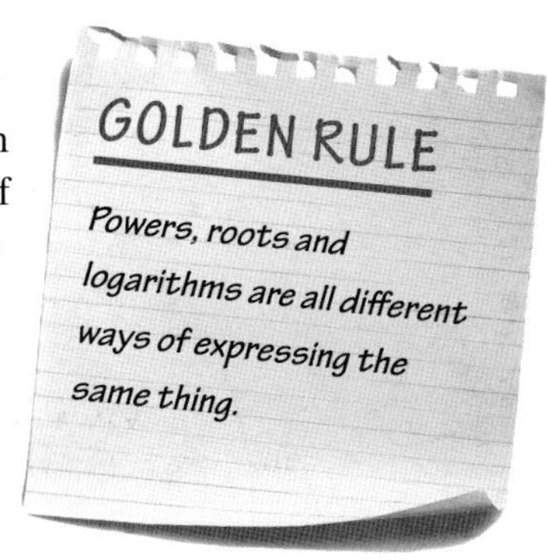

There are some words in mathematics which strike fear into the soul, conjuring up the image of something unimaginably technical and incomprehensible. One such culprit is the word 'logarithm'. But, in truth, logarithms (or 'logs' to their friends) are much tamer creatures than their fearsome reputation suggests. They are just the opposites of powers, in the same way that subtraction is the opposite of addition, and division is the opposite of multiplication.

But how can this be true? Aren't roots the opposite of powers, as we have just seen? Yes they are! And yet roots and logs are not the same thing. In fact, the best picture is to see powers, roots and logarithms as three corners of a triangle.

To answer the question $2^? = 8$ is to reason about logarithms. In this case the answer is 3, and we say that 3 is 'the logarithm of 8 to base 2'. This is written as '$\log_2 8 = 3$'. Although this looks complicated, the meaning of this expression is exactly the same as '$2^3 = 8$'.

The key to answering questions about logarithms is to translate them into the more familiar 'powers' notation. So, when faced with a challenge such as to find $\log_3 9$, the first step is to translate it into a more comfortable form: $3^? = 9$. So the question 'find $\log_3 9$' just means 'how many times do we need to multiply 3 by itself to get 9'. When translating between logarithms and powers, the base of the logarithm (that's 3 in this example) is the number which gets raised to a power. The challenge is to figure out what the power is.

TRY QUIZ 4 – YOU DON'T NEED A CALCULATOR!

As you might expect, many questions about logs need a calculator. But beware! Some calculators have *several* buttons related to logs, and some have none. A particular warning is that the button simply marked [log] usually means 'log to base 10', that is, $\log_{10}$.

The general log button is likely to be marked [$\log_x y$] or [$\log_{\square}\square$]. However, not all calculators have this button; on some calculators you cannot calculate general logarithms directly, only those to base 10.

So powers, roots and logarithms are all different perspectives on the same idea, namely expressions involving three numbers, such as $2^3 = 8$. Whether we want to use a power, a root or a log depends on which two numbers are given, and which is left to be calculated.

- You might be asked '$2^3 = ?$'. This is a straightforward question about powers.
- If you need to answer '$?^3 = 8$' this is a question about roots, and has the same meaning as $\sqrt[3]{8} = ?$
- If you are faced with '$2^? = 8$', that is to say, '$\log_2 8 = ?$', this is a question about logarithms.

We can put these three possibilities in a table. Whether we want a power, a root or a log is a question of which two of the numbers are given, and which is left to be calculated.

NAME	NOTATION	WHAT YOU HAVE TO CALCULATE
Powers	2^3	$2^3 = ?$
Roots	$\sqrt[3]{8}$ or $8^{\frac{1}{3}}$	$?^3 = 8$
Logs	$\log_2 8$	$2^? = 8$

The usefulness of logarithms

There is one fact about logarithms which made them very useful before the invention of the calculator: when you multiply two numbers together, and then take the logarithm, this is the same as adding the two logarithms of the original numbers together.

To put it another way:

$\log(x \times y) = \log x + \log y$ (*The law of logarithms*)

This is true for any two numbers x and y. I have left the base off the logarithms here because it doesn't matter what it is, so long as the three logarithms have the same base.

This means that when we are faced with something like $\log_3 11 + \log_3 2$, rather than working out the two logarithms separately, we can immediately combine them into a single calculation: $\log_3 22$

This prompts two questions: why should this be true? And who cares?

The reason it is true follows from the first law of powers which we met in an earlier chapter. This says that for any three numbers $a^b \times a^c = a^{b+c}$. If we take logarithms to base a, we get the law of logarithms.

(For the more ambitious reader, here is the argument: If $x = a^b$ and $y = a^c$, then, taking logarithms to base a, it follows that $b = \log x$ and $c = \log y$. Also, we know from the first law of powers that $x \times y = a^{b+c}$. Taking logarithms of this, we get $\log(x \times y) = b + c$, which says that $\log(x \times y) = \log x + \log y$.

ENJOY WORKING WITH LOGARITHMS IN QUIZ 5!

The mathematics of years gone by

The law of logarithms has been remarkably important in the history of science and technology. The reason is that it converts questions of multiplication (which are potentially very tricky), into questions of addition (which are much easier). Before pocket calculators, the standard piece of mathematical equipment was a book of log tables, which listed the logarithms of lots of numbers, to a fixed base (such as 10).

To multiply two large numbers such as 187 and 2012, the procedure was as follows: look up the logarithm of each number in the book of log tables. (Any base will do, so long as we are consistent with our choice. Let's take base 10.) These numbers have logarithms 2.27184 and 3.30363, respectively (to five decimal places). Next we add these to get 5.57547. To find the final answer, 'undo' the logarithm (that is $10^{5.57547}$) by looking it up in the log tables to find the number that has this as a logarithm. This gives a final answer of 376,244. You can check that this is correct!

Sum up *Powers, roots and logarithms are close cousins. If you understand one, you understand them all, so long as you can remember how they are related!*

Quizzes

1 Find these square roots.

a $\sqrt{4}$

b $\sqrt{100}$

c $\sqrt{64}$

d $\sqrt{49}$

e $\sqrt{144}$

2 Use a calculator to work these out to two decimal places.

a $\sqrt{10}$

b $\sqrt[3]{10}$

c $\sqrt[4]{10}$

d $\sqrt[5]{56}$

e $\sqrt[7]{2.87}$

3 Work out these fractional powers.

a $121^{\frac{1}{2}}$

b $16^{\frac{1}{4}}$

c $16^{\frac{3}{4}}$

d $25^{\frac{3}{2}}$

e $27^{\frac{4}{3}}$

4 Find these logarithms.

a $\log_6 36$

b $\log_3 81$

c $\log_4 64$

d $\log_2 128$

e $\log_5 125$

5 Simplify these logarithms (to get a single answer such as $\log_{11} 12$).

a $\log_4 6 + \log_4 8$

b $\log_3 13 + \log_3 2$

c $\log_{10} 9 + \log_{10} 8$

d $\log_5 7 + \log_5 6$

e $\log_6 11 + \log_6 10$

Percentages and proportions

- *Understanding proportions*
- *Translating between percentages, decimals, fractions and ratios*
- *Calculating percentages of quantities*

Algebra

- *Realizing what it means when letters appear in equations*
- *Understanding how algebra can represent real-life situations*
- *Learning the rules for working with algebra*

4 Find the percentage increase or decrease.

a The number of people over 100 years old was 40 last year and 42 this year.

b The number of cars in my road was 31 last year and 38 this year.

c A girl is 0.74 metres tall one January, and 0.98 metres the same time next year.

d The number of DVDs sold is 13,488 in December and 11,071 in January.

e Cases of measles rose from 419 to 1012.

5 Work out the quantities for these recipes.

a Bread dough weighing 1kg is made from flour and water in the ratio 5:1.

b An exercise regime tells you to go 3 miles, jogging and walking in the ratio of 4:2.

c Cake mixture weighing 500 grams is made from eggs, butter and flour in the ratio 1:3:6.

d A drink of 360ml is mixed from fruit cordial and water in the ratio 1:8.

e A Spanish omelette weighing 440g is made from eggs, onions and potatoes in the ratio 5:2:4.

6 How much money do you have in each scenario?

a You put $200 in an account at 3% interest, for 5 years.

b You put $50 in an account at 4% interest, for 15 years.

c You put $3 in an account at 20% interest, for 10 years.

d You put $1000 in an account at 3.5% interest, for 30 years.

Every time you open a newspaper, or watch the news on TV, there is a certain type of number that you are guaranteed to see. This is the* percentage, *indicated by its own little symbol '%'. What is a 'per cent'? It is a hundredth, nothing more, nothing less. To say that you have eaten 75% of a sandwich is exactly the same thing as saying that you have eaten 75 hundredths of it.

Percentages act as a convenient scale for measuring how much of something you have, with 0% meaning none of it, and 100% meaning all of it. Half-way along is the 50% mark.

We have already met another way of writing hundredths, namely decimals (see *Decimals*). The percentage 13% stands for thirteen hundredths, or $\frac{13}{100}$, and the decimal 0.13 means exactly the same thing.

As this example suggests, translating between percentages and decimals is straightforward. It is just a question of multiplying (or dividing) by 100, which simply means shifting the digits by two steps, relative to the decimal point. (Have a look at the chapter on decimals for a reminder of how to this works.)

So 99% is the same as 0.99, while 76% is the same thing as 0.76, and 8% is the same thing as 0.08. The last of these is the only one where a mistake might be made; you might think that 8% is the same thing as 0.8, rather than 0.08. But remember that the first column after the decimal point is the tenths column, and the second one is the hundredths column. As 8% means exactly '8 hundredths', it must be equal to 0.08. On the other hand 0.8 represents the same thing as 80%.

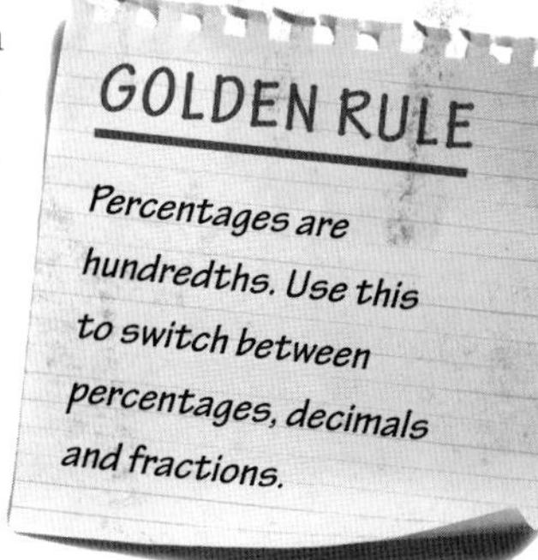

The same system works even when the percentage itself has some decimal places, as in 64.3%. To change this to a decimal is simply a matter of moving the figures two steps to the right, relative to the decimal point: 0.643.

Calculating percentages: from percentages to decimals

How much is 75% of a sandwich? Suppose your sandwich originally weighs 100 grams. Then it is easy to work out how much 75% is: 1% is the same as one hundredth of the sandwich, that is, 1 gram. So 75% is just 75 grams. If the sandwich was 200 grams, then 1% would be 2 grams, and 75% would be 150 grams.

Usually, of course, the figures are less neat than this. What if we want to work out 13% of a 267 gram sandwich?

To calculate 13% of 267, the first step is to translate 13% into a decimal: 0.13. The second step is then to *multiply* 267 by 0.13 (remembering that 'of' means multiply). If we [illegible] a calculator, this is easy enough (otherwise we can [illegible] od for doing it by hand; see *Decimals*).

NOW CALCULATE SOME PERCENTAGES IN QUIZ 1.

Some calculators [illegible] special [%] button. How does this work? Not very w[illegible] experience! I would advise against using it.

Proportion: from fractions to percentages

Percentages are just one way of measuring *proportion*, that is, the relative sizes of two quantities. We have already met two other ways: decimals and fractions. If there are 25 people in a room, of whom 13 are female, what percentage does this represent? It is easy to represent this as a fraction: $\frac{13}{25}$ of the people in the room are female. To convert this to a decimal, multiply top and bottom by 4 to get an equivalent fraction expressed in hundredths (see *Fract[illegible]*). This gives $\frac{52}{100} = 0.52$. The final step, then, is to convert this to a percentage by moving the digits two places with respect to the decimal point: 52%.

IF THAT SEEMS STRAIGHT FORWARD, THEN TRY QUIZZES 2 AND 3.

It is important to be able to translate between the three languages of percentages, fractions and decimals.

Percentage increase and decrease

Some of the most misunderstood statistics have to do with percentage increase and decrease.

Suppose that some new houses get built on my street, and 50 new people move in, raising the number of residents from 200 to 250. What is the 'percentage increase'? Well, the increase clearly amounts to 50 people. So, to answer the question, we need to know what 50 is as a percentage of 200. Notice that it is 200, not 250, here: the increase is always gauged against the starting value. Well, 50 is one quarter of 200, which is 25%. So the population of the street increased by 25%.

The rule here is:

> *Percentage increase or decrease is the difference between the old and new numbers, as a percentage of the old number.*

There is a warning which accompanies all these types of figures: they can feed into a brand of sensationalism beloved of newspaper headline writers.

Suppose your newspaper declares that 'Ca [illegible] Jill's disease up 400% this month!'. This sounds like a catastrophe i [illegible] njuring up images of an epidemic exploding out of contro[illegible] Bu [illegible] figures might be rather more prosaic. It might be that one [illegible] ıt the disease last month, and five caught it this month. Tha [illegible] nds to a 400% increase, and yet might be within the usual range o [illegible] on. If the average (that is to say *mean*, see *Statistics*) monthly figures are [illegible] e cases per month, then last month was a little low, and this month is a little high without being a major cause for concern. Of course, taken across the population as whole (around 60 million in the UK), the risk remains vanishingly small. The lesson here is:

Warning! Percentage increases and decreases provide *no information whatsoever* about how widespread a phenomenon is!

They are not even particularly useful in describing how the figures are changing, since an increase of 1000 cases might represent an increase of 200% (if the numbers have risen from 500 to 1500), or equally an increase of 0.1% (if the numbers have risen from 1,000,000 to 1,001,000). For these reasons, statisticians often complain about the overuse of 'percentage increase', and similar statistics. It would be more informative to report the actual figures, rather than sensational statistics.

NOW TRY QUIZ 4.

Ratios

Percentages provide one way to express proportions. Decimals and fractions are other possibilities. Now we will have a look at yet another.

Suppose I am mixing a cocktail, and a recipe tells me to mix orange juice and vodka in the *ratio* of 3:1. How do I know how much of each ingredient to pour? In particular, suppose I want the final drink to fill a 200ml glass. How much orange juice and vodka do I need?

A recipe might express this as combining 'three parts orange juice with one part vodka'. This way of expressing the recipe gives a clue how to calculate the quantities. The crucial observation is that the mixed drink will consist of *four* parts in total.

Once we realize this, finishing off is fairly easy. The four parts in total should amount to 200ml. So *one* part must be 200ml $\div$ 4 $=$ 50ml. We want three parts of orange juice, which means 3 $\times$ 50ml $=$ 150ml, and one part of vodka, meaning exactly 50ml.

Why did the recipe use the language of ratios instead of just saying 150ml juice and 50ml vodka? The answer is that ratios can easily be scaled up. If instead we wanted to mix a 2 litre jug of cocktail, the same ratio 3:1 remains valid, even though the exact quantities change. This time one part becomes 0.5 litre, so the final mixture is 1.5 litres of juice, and 0.5 litre of vodka.

Let's apply the same line of reasoning to a more complicated ratio. Suppose another cocktail requires tomato juice, lemon juice and vodka in the ratio of 5:1:2. This time there are going to be eight parts in total, because 5 $+$ 1 $+$ 2 $=$ 8. We could express the recipe as fractions: it should be $\frac{5}{8}$ tomato juice, $\frac{1}{8}$ lemon juice, and $\frac{2}{8}$ vodka.

To convert a ratio into fractions, the rule is:

> *Add up the total number of parts in the ratio. This goes on the bottom of all the fractions, and the original numbers go on the top.*

HAVE A GO AT WORKING OUT SOME RECIPES YOURSELF IN QUIZ 5.

If we then want to convert ratios into exact measurements, we multiply these fractions by the total quantity required. So if we want 2 litres of the tomato–lemon–vodka cocktail, we need to multiply 2 by each of the fractions $\frac{5}{8}$, $\frac{1}{8}$ and $\frac{2}{8}$, in turn, to get measurements of 1.25 litres tomato juice, 0.25 litres lemon juice, and 0.5 litre vodka.

Interest rates

One place that percentages commonly occur is as interest rates. Suppose you put \$100 in a bank account that has an annual interest of 5%. Since 5% of \$100 is \$5, after one year, your account should have \$105 in it.

What about after another year? A common mistake is to believe that, each year, the amount will grow by \$5, producing \$110 after two years. But the amount grows by 5%, not by \$5. In the second year, it will grow by 5% of what was there at the beginning of the year, that is, \$105. To work out 5% of \$105 calculate $105 \times 0.05 = 5.25$. So, after two years, the account will contain \$110.25. What about after 10 years?

To find the total at the end of a year, we first multiplied the amount of money at the start of the year by 0.05 and then added the result to the starting figure. A shortcut is to multiply the figure at the start of the year by 1.05. So, after one year, the account contains 100×1.05. After two years, it contains $100 \times 1.05 \times 1.05$, that is to say 100×1.05^2. After three years, it contains 100×1.05^3, and so on. (Look at *Powers* if you need to remind yourself how these work.) So, to answer the question above, after 10 years, the account contains 100×1.05^{10}, which comes out at around \$162.89. (If the amount grew by \$5 per year, after 10 years the account would contain only \$150, so the difference is significant.)

In general, to calculate how much money an account contains, we need three pieces of data. First we need the original deposit made into the account. Call this *M*. Then we need to know the interest rate of the account, making sure this is expressed as a decimal larger than 1 rather than a percentage. (An interest rate of 5% or 0.05 would be expressed as 1.05; this means the original amount plus interest.) Call this number *R*. Finally we need to know how many years ago the deposit was made. Call this *Y*. The formula for the amount of money currently in the account is:

$$M \times R^Y$$

Of course, if tax is deducted, the bank changes its rate, or money is moved in or out, then the story becomes more complicated! Nevertheless, this formula captures the fundamental rule of interest rates.

IT'S TIME TO TRY THIS OUT FOR YOURSELF. HAVE A GO AT QUIZ 6.

Sum up *A decimal and a percentage are almost the same thing; you just need to multiply the decimal by 100. In the same way, ratios and fractions are almost the same thing. Beware: percentage increase or decrease does not tell us anything about the frequency of a phenomenon!*

Quizzes

1 Work out each of these percentages as a number.

a Of a pack of 15 dogs, 20% are spaniels.
b In a factory of 400 workers, 77% work full-time.
c In a town there are 1225 roads, of which 32% are designated no-parking.
d In a colony of 134,550 ants, 88% are workers.
e A human brain contains 160 billion cells, of which 54% are neurons.

2 Express each of these proportions as a percentage, to the nearest 1%.

a A packet contained 22 sweets of which 7 are left. What proportion are left?
b An office has 86 workers, of whom 14 are off sick. What proportion are off sick?
c 197 of the 458 houses in a village are thatched. What proportion are thatched?
d In a city of 732,577 people, 118,504 are children. What proportion are children?
e In a country of 8 million people, 0.5 million wear glasses. What proportion wear glasses?

3 Convert these percentages to both decimals and fractions.

a 20% **b** 100% **c** 99% **d** 5% **e** 4%

For many people, the moment when mathematics moves from being fairly simple to being incomprehensible is when letters start appearing where previously there were only numbers. This is* algebra. *In this chapter we will have a look at it. We'll see what it means, and how to do it without getting confused. Most importantly, we will see why it is useful.

Let's start with an example. Suppose a restaurant bill comes to $40, and is to be divided between 8 people. What calculation do we have to do to work out how much each diner pays? The answer is $\frac{40}{8}$. But what if there were only 6 people? Then the answer is $\frac{40}{6}$. And what if the bill was actually $140? Then the calculation is $\frac{140}{6}$. Each of these produces different answers: as the numbers we put in change, so do the numbers we get out.

Yet there is a sense in which they are actually all the *same* calculation. Each time, the total bill is divided by the number of diners. We could write this as:

$$\text{Cost} = \frac{\text{Bill}}{\text{Number of diners}}$$

This has an advantage over the previous versions as it makes explicit what is going on, what *principle* is being applied here. So if the numbers are altered, because of a miscount, or an item being missed from the bill, the same *idea* continues to work.

Here's another example. Suppose I am cooking dinner for a group of people. How many baked potatoes do I need? I reason that each adult diner will eat 2, and each child will eat 1, and that I should have 5 as spares in case anyone wants a second helping. This rule comes out as: Potatoes = 2 × Adults + Children + 5. Then, when the numbers of guests have been clarified, I can put this principle into action. Once I know that there will be 5 adults and 3 children, I can plug these numbers into my rule to arrive at $2 \times 5 + 3 + 5 = 18$ potatoes.

The fun of formulae

The discussion so far gives us the idea of algebraic *formulae*. Even if you wouldn't usually write these sort of rules down as I have done above, I hope you agree that this type of thinking is quite normal and natural.

Well, this is algebra. The only difference when experts do it is that, instead of writing words in their mathematical expressions, they usually cut down to single letters. What is the point of this? To make things neat and tidy, and to save space, of course! (It has not always been thus: mathematicians of bygone eras often wrote lengthy prose in amongst their equations.)

GOLDEN RULE

Instead of calculating with individual numbers, use algebra to express rules that work for any numbers.

So, in the potato calculation above, I might begin by calling the number of adults a and the number of children c. Then the number of potatoes I need to cook (call it p) must satisfy:

$$p = 2 \times a + c + 5$$

It is usual to omit the $\times$ signs when writing algebra using letters, so we would write this as:

$$p = 2a + c + 5$$

What we have arrived at is a typical example of an *equation*, or a *formula*. The power of this method is that it expresses lots of different facts in just one line.

Many mathematical facts are expressed in this sort of way. For example, the area of a rectangle is expressed by multiplying its length by its width. We might write this rule as $A = l \times w$ (where A, l and w stand for the area, length and width, respectively).

The ability to translate between algebraic formulae and English sentences is one of the central planks of mathematical thinking, and well worth spending some time on.

TURN STATEMENTS INTO ALGEBRAIC FORMULAE IN QUIZ 1.

From numbers to letters and back: substitution

We have seen how to turn English sentences into mathematical formulae. What can we do with these formulae? When all is said and done, we are probably hoping for a number at the end of the calculation, rather than a collection of letters and algebraic symbols.

To extract a number from a formula, we first need to know how to feed numbers into it. If we have the formula $p = 2a + c + 5$, and we are further told that $a = 5$ and $c = 3$, then we can replace the symbols a and c with these new values, and then work out the value of p:

$$p = 2 \times 5 + 3 + 5 = 18$$

What we have done here is to *substitute* numerical values for some of the letters, and then work out the final answer.

We also saw above that a rectangle's area is given by the formula $A = l \times w$. If a particular rectangle has values of $l = 8\text{cm}$ and $w = 3\text{cm}$, then we can substitute these values into the formula to get an area of $A = 8\text{cm} \times 3\text{cm} = 24\text{cm}^2$.

The ability to substitute values into formulae becomes more and more important in all branches of the subject, as the mathematics becomes more complex. You might object to the previous examples by saying 'multiply the length by width' is quick and simple enough, and doesn't really need to be abbreviated as a formula. But if we want to calculate the volume of a cone (see *Area and volume*), the formula '$V = \frac{1}{3}\pi r^2 h$' is a lot more concise (and, with practice, easier to read) than writing 'to find the volume, multiply the radius of the base circle by itself, and then by the length of the cone, then divide by 3, and multiply by the ratio of a circle's circumference to its diameter'.

HAVE A GO AT SUBSTITUTING VALUES INTO FORMULAE IN QUIZ 2.

Tidying up algebra

There are various rules that we can use to make formulae simpler. (These will be invaluable when we come to solve equations later.)

The idea is very familiar, when expressed in terms of numbers: just as we can add up $2 + 3 = 5$, similarly we can add $2x + 3x = 5x$ and $2a + 3a = 5a$ when letters are involved.

Why should this be so? Think of a number and double it. Then add on your original number tripled. The answer is five times your original number. Magic! Hardly. This will always work, irrespective of what number you choose, and this is the rule expressed by $2x + 3x = 5x$. The x, as we have seen, is standing for any number.

This rule is useful for tidying up, or *simplifying*, algebra. If we have an expression such as:

$$2 + 3x + 5x + 2 + 2x$$

then it can be simplified by collecting together the plain numbers: $2 + 2 = 4$ and collecting together the xs: $3x + 5x + 2x = 10x$, to leave us with a much tidier expression: $4 + 10x$.

The same thing works when there are more letters involved. If we are presented with $a + 4 + 2b - 5 + b + 3a$, then we can gather the plain numbers together: $4 - 5 = -1$, and the as: $a + 3a = 4a$ and the bs: $2b + b = 3b$, giving a result of $4a + 3b - 1$.

Warning! Simplifying algebra is always a good idea, where possible. But one of the commonest mistakes is to try to simplify things where it cannot be done. For example, while $b + 2b$ can be simplified to $3b$, if we are faced with the expression $b + b^2$, there is no way to simplify this. It is not equal to $2b$ or $2b^2$. (Why not? Well, if $b = 10$, then $b + b^2 = 110$, while $2b = 20$, and $2b^2 = 200$.) Similarly if we have $a + b + ab$, this cannot be simplified, and should be left as it is.

SIMPLIFY SOME ALGEBRA IN QUIZ 3.

Algebra and brackets

Here's a trick: Think of a number, any number! Now add 4, and then double what you get. Now add 2. Next, halve the result, and then subtract the number you first thought of. And the answer is . . . 5. Alakazam!

How does this work? It is a simple consequence of the algebra of brackets, which is what we are going to look at in the final section if this chapter. We'll see a detailed explanation later on!

Brackets are useful for avoiding ambiguity when writing out calculations (see *The language of mathematics*). But they are even more important when algebra is involved.

The key insight is this. Suppose I add 3 to 5 and then double the answer. We might write this as $2 \times (3 + 5)$. It is no coincidence that this comes out the same as doubling 3 and 5 individually, and then adding together the two results: $2 \times 3 + 2 \times 5$.

In fact, this is exactly the principle used for doing long multiplication: that $10 \times (50 + 2)$ is the same as $10 \times 50 + 10 \times 2$. We call this *expanding brackets*. The idea is as follows: when you have something being added (or subtracted) inside a pair of brackets, and something outside the brackets multiplying (or dividing) the brackets, this is the same as performing the multiplications (or divisions) individually, and then adding up the answers.

In algebra, we might write $a \times (b + c) = a \times b + a \times c$. Using the convention of omitting multiplication signs, this becomes $a(b + c) = ab + ac$. The great thing is that this is true whatever a, b and c are.

So, if we are faced with $2(x + 3y)$, we expand the brackets to get $2x + 6y$. Similarly $x(x - 3y) = x^2 - 3xy$. These are both just special cases of the general rule.

Let's go back to the trick we started the section with, and let's call the mystery number x. The first instruction is to add 4 to it, giving $x + 4$. Doubling that produces $2(x + 4)$. At this stage, let's expand this brackets: $2x + 8$. Adding on another 2 gives us $2x + 10$. Next we were told to halve the result, which we can write as $\frac{1}{2}(2x + 10)$, and again, let's expand the brackets, producing $x + 5$ The final instruction was to subtract the number we first thought of, which of course is x. But now it is as clear as day that subtracting x from $x + 5$ will always leave us with 5. It's not so much *Alakazam* as *Algebra!*

EXPLORE THE ALGEBRA OF BRACKETS IN QUIZ 4.

Why not try coming up with some of your own tricks along these lines?

Sum up *Algebra is a great language for expressing general rules and laws. Just remember how to translate between algebra and English!*

Quizzes

1 Turn these statements into algebra.

a The number of animals on the ark is twice the number of species on Earth. (Let a be the number of animals on the ark and s be the number of species on Earth.)

b The amount of cake on my plate (c) is two divided by the number of people present (p).

c The number of hours to cook the meat (h) is one quarter of its weight in pounds (w) plus an extra half-hour.

d The number of patients in the hospital (p) is four times the number of doctors (d) plus the number of wards (w).

e The temperature in Fahrenheit (F) is the temperature in Celsius (C), multiplied by nine, divided by five, and then with thirty-two added.

2 Substitute the values into the formulae.

a If $B = t - s$, then what is B when $t = 13$ and $s = 5$?

b If $x = \frac{y}{54}$, then what is x when $y = 108$?

c If $a = \frac{2}{b^2}$, then what is a when $b = 4$?

d If $D = y + \sqrt{z}$, what is D when $y = 10$ and $z = 16$

e If $z = x^2y$, what is z when $x = 6$ and $y = 2$?

3 Simplify these.

a $a + 3a$

b $b + 5 + 2b - 4$

c $x + 4y + 2x - 2y$

d $5x + 5 + x - 3a - 5$

e $x + 3z + 2y + 2z + 2$

4 Expand these brackets.

a $4(x + z)$

b $2(x + 4)$

c $x(x - 1)$

d $x(x - 2y)$

e $2x(x - 2y)$

Equations

- *Knowing what equations are*
- *Understanding what it means to* solve *an equation*
- *Getting to grips with techniques for solving equations*

One of the good things about equations is that you do not need anyone to tell you whether you have got it right. It is easy to check, and I would suggest that you always do this. If our solution above is correct, then if we take the original equation $4x - 3 = 17$, and substitute in the value $x = 5$, we should get a true statement. If we don't, then we know we have made a mistake. So let's try: $4 \times 5 - 3 = 17$; this is true, so we have solved the equation correctly.

If we were presented directly with the equation $4x - 3 = 17$, how would we know where to start: which move needs to be undone first? The two options are multiplication by 4 and subtraction by 3. An old friend comes to our assistance: BEDMAS (see *The language of mathematics*). That tells us that the multiplication was done first, followed by the subtraction. So, to undo these steps, we go in reverse order and tackle the -3 first.

The *x*s stick together: collecting like terms

Here is another think-of-a-number problem: I think of a number, treble it and subtract 2. The answer I get is the same as if I double my original number and add 4.

As usual, let's call the mystery number x. If we treble it, we get $3x$, then subtracting 2 gives $3x - 2$. Now the problem tells us that this is the same as another quantity, which we get if we double x to get $2x$, and add 4 to get $2x + 4$. So the equation which encapsulates this problem is:

$$3x - 2 = 2x + 4$$

Now, as always, we are aiming for a final equation of the form $x = \square$. But, this time, things look trickier, as there are *x*s and numbers on both sides of the equals sign. The first step, then, is to improve this situation. We want to eliminate the *x*s on one side or the other.

The tactic is to collect together the *x*s on one side, and the numbers on the other. But how can we do this? Let's deal with the *x*s first, and let's decide to collect them on the left (we could equally well choose the right). That means getting rid of all the *x*s on the right. Well, there are two of them, that is to say $2x$. We can eliminate these by subtracting $2x$ from the right. Of course, to keep the equation balanced, we also have to subtract $2x$ from the left:

$$3x - 2 - 2x = 2x + 4 - 2x$$

(It is not strictly necessary to write this step out; I am just doing it to make explicit what is happening.) Now on the right-hand side, the $2x$ and the $-2x$ cancel each other out as planned, leaving only 4. On the left-hand side, we start with $3x$ and take away $2x$, leaving just x. So our equation now reads:

TRY COLLECTING LIKE TERMS IN QUIZ 3.

$$x - 2 = 4$$

This is much simpler already, and now very quick to finish off. By adding 2 to both sides of the equation, we get the solution $x = 6$.

Let's review that last example: we collected together all the xs on one side, and all of the plain numbers on the other. This technique is called *collecting like terms*, and it is the best way to make a complicated equation simpler.

There are two other little tricks we might need. Firstly, if $A = B$ then it is equally true that $B = A$. This means we can swap over the two sides of an equation any time we like. So if we find that $12 = 3x$, we can easily swap the sides over and write this as $3x = 12$. The second trick involves flipping $+/-$ signs. We can do this so long as we follow the golden rule and do it to both sides. So if we have $-4x = -8$, we can flip the signs to get $4x = 8$. Similarly, if we have $-5x = 10$, we can flip the signs on both sides to get $5x = -10$.

NOW TRY QUIZ 4

Some things are more equal than others: inequalities

The equation has a lesser-known cousin: the *inequality*. Here the nexus is not the equals sign (=), but one of the four inequality symbols. The first two are $<$ and $>$, standing for 'is less than' and 'is greater than', respectively. So we might truthfully write $4 < 7$, or write $x > 9$ to indicate that x is some number larger than 9. These two symbols represent *strict* inequalities. (Alternatively you can think of these two symbols as being just one reversible symbol: $4 < 7$ means the same as $7 > 4$.)

The *weak* inequality symbols $\leq$ and $\geq$ stand for 'is at most' and 'is at least', or 'is less than or equal to' and 'is greater than or equal to', respectively. So, while it is true that $4 \leq 7$, it is also true that $4 \leq 4$. (But it is not true, that $4 < 4$.)

With these symbols in place, we can write inequalities in the same way as we write equations. For example: $5x - 4 < 3x + 2$. But what does it really mean to 'solve' an inequality? We cannot hope for a unique answer. Instead, we want to pin down a range of values of x for which the original inequality is true.

If we follow the rules above, we first collect the xs on the left, by subtracting $3x$ from both sides: $2x - 4 < 2$. Next we want to collect the plain numbers on the right, by adding 4 to both sides: $2x < 6$. Finally we can divide by 2 to get out answer: $x < 3$. This is the solution to the inequality, and expresses the full range of values of x for which the original inequality is true. If $x < 3$ then it should be true, but if $x \geq 3$, it should not. Test it out!

Changing signs and symbols

Solving inequalities is *almost* identical to solving equations. But there is a danger-point where the two diverge. If we find that $8 = 2x$, then we can swap the sides of the equation to get $2x = 8$. With an inequality, however, when we swap sides, we also have to reverse the inequality symbol. So $8 > 2x$ becomes $2x < 8$.

There is another situation where the inequality has to be reversed. In the context of an equation, if we find that $-x = -7$, for example, then we can flip signs on both sides to positive, to get a final answer of $x = 7$. Let's think about the inequality $-x < -7$. The value $x = 8$ does satisfy this, since $-x = -8$, which is indeed less than -7. Similarly the value $x = 6$ does not satisfy $-x < -7$, since -6 is not less than -7. If we simply flip the sign on both sides of the inequality $-x < -7$ (as we do with equations), we get $x < 7$. But we have just seen that this is not the right answer. The rule is that when you change the signs in an inequality (or, equivalently, when you multiply or divide both sides by a negative number) you need to reverse the inequality symbol. So, if we have $-x < -7$, we change the signs to positive on both sides, but in doing so we must also reverse the $<$ symbol, giving an answer of $x > 7$.

SOLVE SOME INEQUALITIES IN QUIZ 5.

Sum up *Solving equations occupies pride of place at the heart of mathematics, as this is the main tool for getting answers to algebraic questions. Just remember the golden rule: keep the equation balanced!*

Quizzes

1 Solve these think-of-a-number problems!

a I think of a number, multiply it by 3. The answer is 18. What was my number?

b I think of a number, multiply it by 4, and then subtract 3. The answer is 25. What was my number?

c I think of a number, divide by 2, and then multiply it by 3. The answer is 9. What was my number?

d I think of a number, subtract 4, and then multiply it by 5. The answer is 20. What was my number?

e I think of a number, subtract 3, and then divide by 5. The answer is 5. What was my number?

2 Solve the problems in quiz 1 again, using algebra.

3 Solve these equations.

a $4x = 2x + 6$ **b** $3x - 4 = 2x + 8$ **c** $5x - 4 = 14 - x$

d $2x + 2 = 12 - 3x$ **e** $1 - x = 8 - 2x$

4 Solve these equations.

a $2(x + 1) = 3x$

b $3(x - 1) = 2(x + 1)$

c $4(x - 2) + 2 = 2(x + 5)$

d $4(2 - x) - 1 = 2(x + 3) + 1$

e $2(1 - x) + x = 3(2 - x) + 2$

5 In quiz 3, try replacing the '=' sign with '<' and solving the resulting inequalites.

Angles

- *Measuring angles*
- *Knowing the rules for calculating angles*
- *Understanding the geometry of parallel lines*

Triangles

- *Recognizing different types of triangle*
- *Reasoning about a triangle's angles*
- *Calculating a triangle's area*

Quizzes

1 Convert these angles from degrees to revolutions.

a 180° **b** 360° **c** 90°
d 270° **e** 540°

2 Use a protractor to draw two lines that meet at each of these angles.
Next to each angle write whether it is acute, obtuse or reflex.

a 45° **b** 72° **c** 83°
d 112° **e** 229°

3 Sketch the following situations and calculate the missing angles.

a Two angles meet at a point. One is 75°. What is the other?

b Three angles meet at a point. One is 110°, another is 40°. What is the third?

c Two angles lie on a straight line. One is 60°. What is the other?

d Three angles lie on a straight line. One is 111°, another is 32°. What is the third?

e Four angles meet at a point. Three are 51°, 85° and 190°. What is the fourth?

4 Sketch the following situations (no need to use a protractor), and calculate all four angles. In each case, two straight lines cross, producing four angles.

a One angle is 90°. What are the other three?

b One angle is 45°. What are the other three?

c One angle is 21°. What are the other three?

d One angle is 122°. What are the other three?

e One angle is 176°. What are the other three?

It is often useful to know when two angles are equal to each other. We have seen some examples above, namely opposite angles. Parallel lines also present situations where equal angles can be found. Let's suppose we have a pair of parallel lines, and a third line which crosses them both (like a straight stick lying across a pair of railway tracks). In such a situation, there are two handy rules which tell us which of the resulting angles are equal.

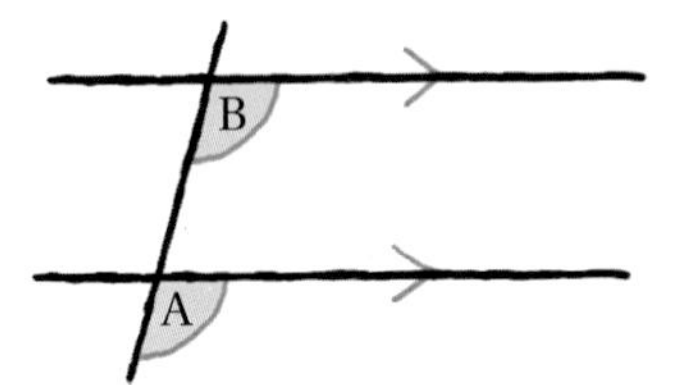

To start with, angles which are in identical positions, but on the two different lines, are said to be in *corresponding* positions. The first rule is:

Angles in corresponding positions are always equal.
(*Law of corresponding angles*)

This should not come as too much of a surprise, since the two angles are identical in every respect, but are just in different places. Sometimes they are known as 'F' angles, because the pattern looks like a capital F. (Don't rely on this, though, as the F may be back to front or upside down, or otherwise deformed!)

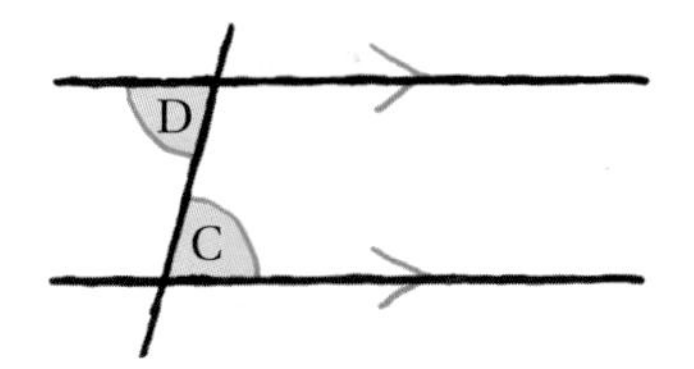

The second law of parallel lines and angles deals with the slightly subtler notion of *alternate angles.* Pairs of alternate angles lie on opposing sides of the third line, and are sometimes known as 'Z' angles.

Alternate angles are always equal.
(Law of alternate angles)

Why should this be true? It is a consequence of the laws on opposite angles, and corresponding angles. Try to see why!

Sum up *Angles and parallel lines are the salt and pepper of geometry. Just remember how to combine them, and they will work wonderfully together!*

angles do not add up to a full 360°. If you turn through one and then through the other, you have completed a *half* turn, which is 180°. Situations like this are known as *angles on a straight line*.

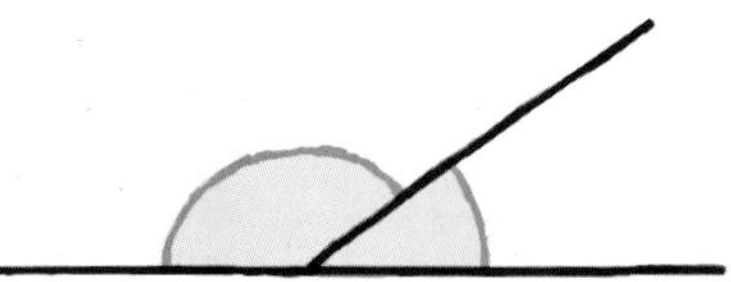

Angles on a straight line add up to 180°.

So as soon as we know one of the angles, say it is 35°, then we can calculate the other: 180° − 35° = 145°.

NOW HAVE A GO AT QUIZ 3.

The attraction of opposites

When two perfectly straight roads cross each other, four angles are formed. Because these are angles at a point, the four must add up to 360°. But there is something else we can say too: if you look at the picture below, the angles seem to come in pairs: two large ones and two small ones.

This is not an illusion, in fact the two small angles must be equal to each other, as must the two large ones. These are known as *opposite angles*, and the rule is:

Opposite angles are always equal.

Why is this true? Well, one large angle together with either of the two small angles form angles on a straight line, and so must add up to 180°. That means that the large angle determines *both* of the small angles, each by exactly the same calculation. As soon as we know just one of the four angles, we can quickly work out all the others.

NOW IT'S TIME FOR QUIZ 4!

Never meeting – parallel lines

The notion of two lines being *parallel* is fundamental. It means that they can be continued for ever, in both directions, without ever crossing. To say the same thing in a different way: the distance between two parallel lines is always the same, no matter where you measure it. Yet another way to say this is that the angle between parallel lines is 0°. In geographers' terminology, they have the same bearing. Train tracks are a classic example of parallel lines.

In geometry, lines are indicated to be parallel by drawing matching arrows on them.

***This chapter contains truly ancient knowledge. Everything we shall see here was included in the greatest textbook of all time: the* Elements, *written by Euclid of Alexandria around 300* BC. *It may seem surprising that Euclid's geometry continues to be studied thousands of years later. But he got to grips with the root ideas of geometry: straight lines and circles, and the ways they interact. We will explore these in later chapters, including the many shapes that can be built from straight lines, of which the most important are triangles. But first we look at the fundamental geometrical notion of an* angle.**

An angle is an amount of turn. An ice skater or ballet dancer who spins on the spot might turn through a large angle. During a long journey, a car tyre is likely to spin through a very large angle indeed! How do we measure angle? Well, just as distance can be measured in miles, centimetres or light-years, there are different units for angle too.

The revolution is coming

A common unit of turn is the *revolution*. If you stand on the spot, make one complete turn and end up facing in exactly the same direction as when you started, then you have completed one revolution. In engineering, the speeds of wheels and cogs are often discussed in terms of *revolutions per minute (rpm)*. The second hand on a clock has a rotational speed of precisely 1rpm, while a washing machine drum might whizz round at 1000rpm.

An even more common unit is the *degree*. This is generally preferable when measuring small quantities of turn, less than a single revolution. One revolution is broken up into 360 degrees (usually written 360°). So if you turn on the spot until you are facing exactly backwards, then you have turned 180° (that is, half of one revolution).

GOLDEN RULE

One complete revolution is 360°.

In many sports, such as skateboarding, ice-skating and diving, turning is an important ingredient. Typically, the amount of turn is measured in degrees. So a skateboarder might say that they have performed an 'Ollie 180'. This means that they have turned 180°, that is $\frac{1}{2}$ a revolution. (What exactly an 'Ollie' may be is a matter for another day!) Similarly a turn of 720° translates to two complete revolutions, because $720 = 2 \times 360°$.

In terms of the numbers, to translate from revolutions to degrees, we need to multiply by 360. So $\frac{1}{4}$ revolution is, in degrees, $\frac{1}{4} \times 360 = 90°$. Going the other way, to convert degrees to revolutions we need to divide by 360. So, as we saw, 720° is $\frac{720}{360} = 2$ revolutions.

GOT ALL THAT? NOW HAVE A GO AT QUIZ 1.

It is useful to practise translating between the language of degrees and revolutions (especially when talking to skateboarders!).

Here is some jargon about angles:

- An angle of 90° (or $\frac{1}{4}$ revolution) is known as a *right angle*. Saying that two lines are *perpendicular* means that they cross at right angles. Also notice the little box we draw at the angle, to indicate a right-angle.
- An angle less than 90° is known as *acute*.
- An angle between 90° and 180° is called *obtuse*
- Angles between 180° and 360° go by the name of *reflex*.

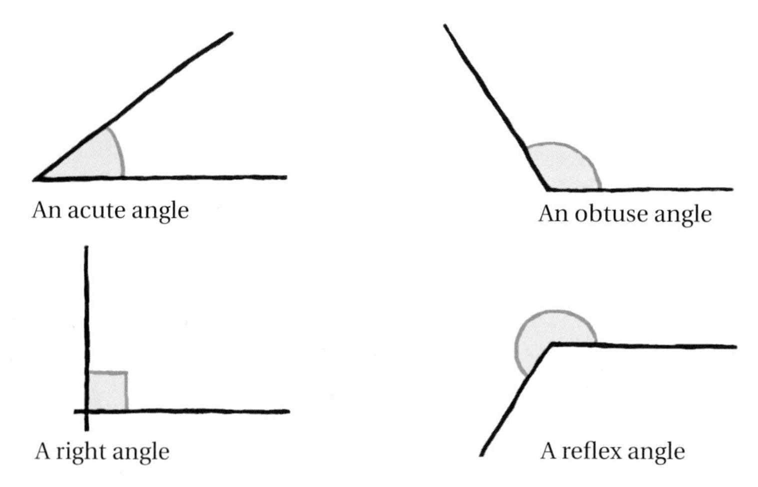

Measuring and drawing angles

There are various pieces of equipment we can use for measuring angles. One is a compass: not the device for drawing circles, but the type for finding North. If you want to know the angle a road makes to the North–South line, a compass will tell you. Compass bearings and angles are essentially the same thing.

In the context of geometry, though, a more familiar (though lower-tech) tool is a *protractor*. It's a semicircular piece of transparent plastic, with various markings on it: there is a base-line running along the flat edge and, at the centre of this, a cross-hair; around the curved edge, angles are marked from 0 to 180°.

How do we use a protractor? If we have two straight lines which meet at a point, we might want to know the angle between them. What does this mean? imagine that the two lines are roads, and you are standing at the point where they meet, looking along one road. What we want to know is the angle through which you have to turn, so that you are looking down the other road (line). This is a task for a protractor! Lay it on top of the page, with its base-line lined up with one of the two lines, and with the cross-hair sitting exactly on the point where the two lines meet.

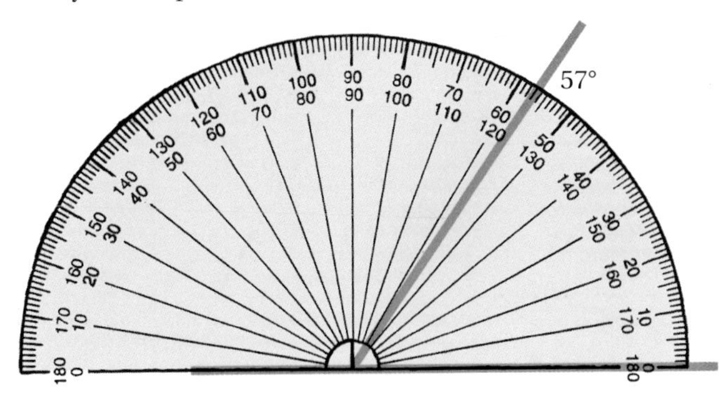

Looking at the curved edge of the protractor, you can see the numbered marks representing angles. The number at the point where the second line emerges is the number we want. Here is a warning though:

Warning! Protractors usually have two scales: one measuring angles clockwise, the other anticlockwise. Make sure you pick the one which counts up from 0°, not the one counting down from 180°!

We can draw angles by adapting this procedure slightly. If we want to draw an angle of 65°, say, the first thing to do is draw one straight line in the position that we want. Then, mark the point on that line where we want the second line to branch off (in this case this is at the end of the line), and place the cross-hair of the protractor on that point, with its base-line matching up with the drawn line. Next, find the 65° on the protractor (bearing in mind the warning above!) and make a mark on the page at that position. Finally, remove the protractor, and join the two marked points with a straight line.

If you have a protractor, try drawing angles in Quiz 2.

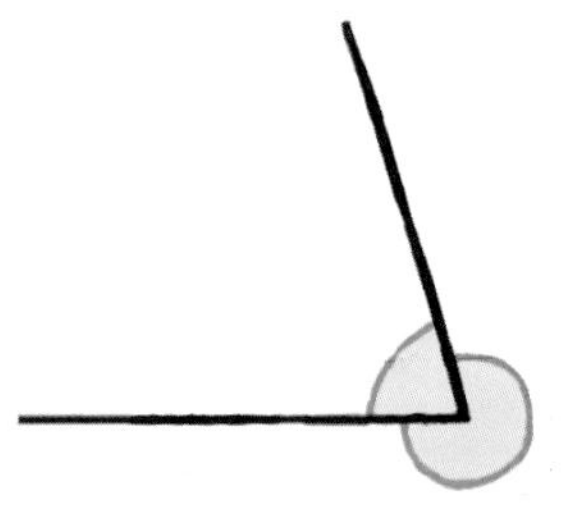

Angles – what's the point?

Look at this picture. It might be showing a road which is perfectly straight except at one point, where it has a corner. Where the two straight sections meet, there are two angles. Imagine that you are standing at the corner, in the middle of the road, facing along one section of road, and then you twist around, so that you are facing down the other. There are two ways you could do this: you could rotate clockwise or anticlockwise, and depending on your choice you will turn through a bigger or a smaller angle.

Now, these two angles are related in a precise way: if you turn through one, and then turn through the other, you will have turned one complete revolution. So, the golden rule tells us that, between them, these two angles add up to exactly 360°. This means that if we know one of them, say it is 72°, then we immediately know that the other must be $360° - 72° = 288°$.

Angles like this are known as *angles at a point,* and this line of thought works just as well when there are more than two involved. We might think of the picture below as a junction of three roads. But again, the three angles must add up to 360°. So if two of them are 37° and 193°, we can immediately calculate the third angle (marked x) as $360° - 37° - 193° = 130°$. No matter how many angles there are, the rule is:

Angles at a point add up to 360°.

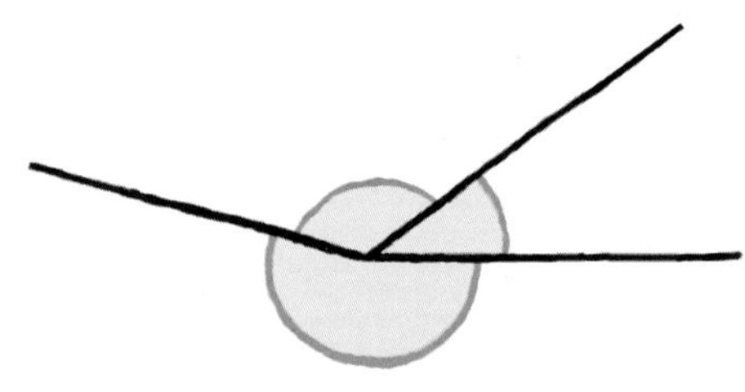

A similar argument works when we have a situation like the one opposite. This time we've got one straight road, maybe a motorway, and a turning coming off it. In this picture, though, the two marked

Two straight lines do not make a shape. But three do. This is why the triangle is one of the most important figures in geometry: it is one of the simplest.

But is the geometry of triangles useful? The answer is yes, as the world is full of triangles, even if most of them are invisible. What, you may ask, are invisible triangles? Whenever you choose three points in space, you have defined a triangle. For instance, I might pick the place where I am standing, where my wife is sitting, and the television in the corner of the room. These three points form a triangle.

In any such situation, the techniques in this chapter can apply, even when the triangle's sides are not immediately obvious. For this reason, triangular geometry really is everywhere, and we shall be meeting more of it later (see *Pythagoras' theorem* and *Trigonometry*).

Different types of triangle

For many people, one triangle is much the same as any other. But, to the connoisseur, triangles come in a variety of forms, each with their own characteristics. The most symmetrical triangle is the one where the three sides are all the same length. Triangles like this are called *equilateral* (coming from the Latin for 'equal sides').

It automatically follows that the three *angles* inside an equilateral triangle must also be equal. As we shall see shortly, each of these angles is equal to 60°. So an equilateral triangle has a very rigid shape, with no room for manoeuvre. The only possible variation is in the triangle's size. In all other respects, every equilateral triangle looks the same, in much the same way that all squares look the same. In fact the equilateral triangle and the square are the first two *regular* shapes, meaning shapes whose sides and angles are all equal. We shall explore this further in *Polygons and solids*.

A slightly less symmetric type of triangle is one which has two of its lengths the same. Such a triangle is known as *isosceles* (coming from the Greek for 'equal legs'). There is more room for manoeuvre with isosceles triangles: they can look very different, as we shall see shortly.

Most triangles are neither equilateral nor isosceles, but have three sides of different lengths. There is a word for this too: *scalene* (from the Greek for 'unequal').

These words – equilateral, isosceles and scalene – refer to the lengths of the triangle's sides. But triangles can also be described by the sizes of their angles. A *right-angled triangle* is, unsurprisingly, one which contains a right-angle. That is to say one of its three angles is equal to 90°. (These are in some ways the 'best' triangles, and we will be hearing a lot more about them in *Pythagoras' theorem* and *Trigonometry*.)

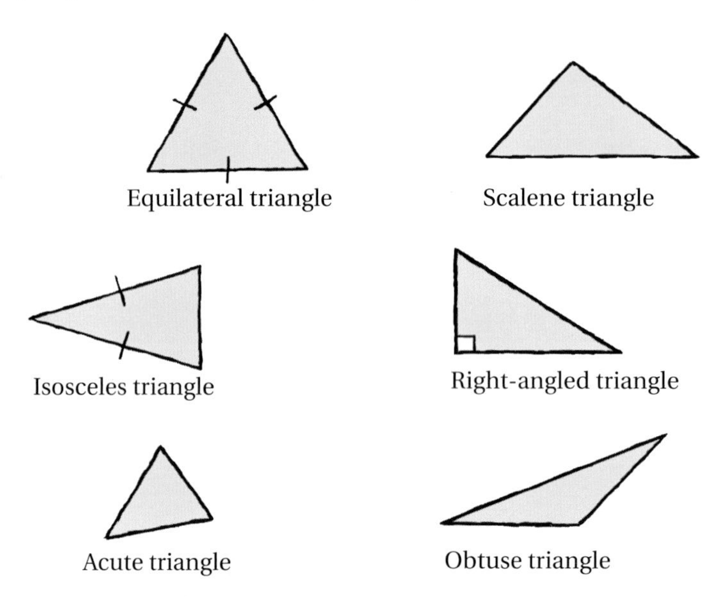

Triangles whose angles are all less than 90° are called *acute* triangles. (Remember from the previous chapter that an angle less than 90° is called an *acute angle*.) Similarly, a triangle which contains an *obtuse angle* (that is to say one that is more than 90°) is known as an *obtuse* triangle.

Got the jargon? Then try quiz 1.

That's a large amount of jargon to digest, and all just to describe different sorts of triangle!

Angles in a triangle

The starting point for the geometry of triangles is a relationship between the three angles inside the triangle. Try drawing a triangle which contains *two* right angles, or *two* obtuse angles. You will fail. Why? The answer is given by this chapter's golden rule.

This rule puts a limit on the possible sizes of the angles inside any triangle.

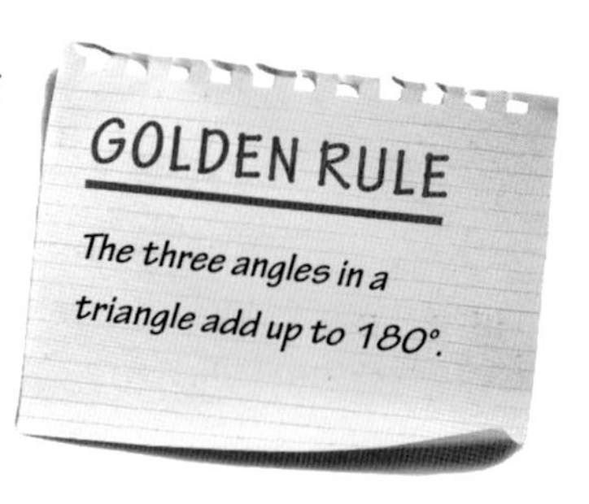

It will prove useful for calculating the sizes of angles in triangles. But why should it be true? It is worth seeing a proof of this famous fact, since it is not complicated. Indeed, it follows from the facts about angles and parallel lines that we saw in the previous chapter.

We begin by taking a triangle, call it ABC, meaning that the three corners are named A, B and C respectively. (These are just labels so that we can refer to them individually.) At each corner is an angle. Since we don't know their values, we had better name these too. Let's call them a, b and c. So what we want to show is that $a + b + c = 180°$.

EXPLORE ANGLES IN A TRIANGLE IN QUIZ 2.

We also have the three lines of this triangle, which we might call AB, BC and CA, where AB is the line running from corner A to corner B, and so on.

The main move in the proof is this: we add in a new line, parallel to AB, which passes through the point C.

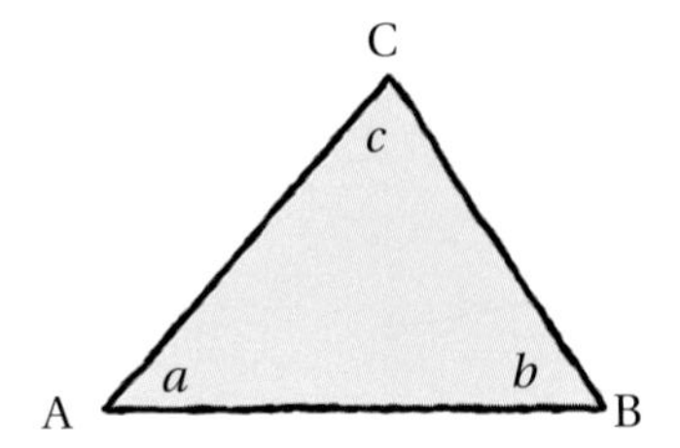

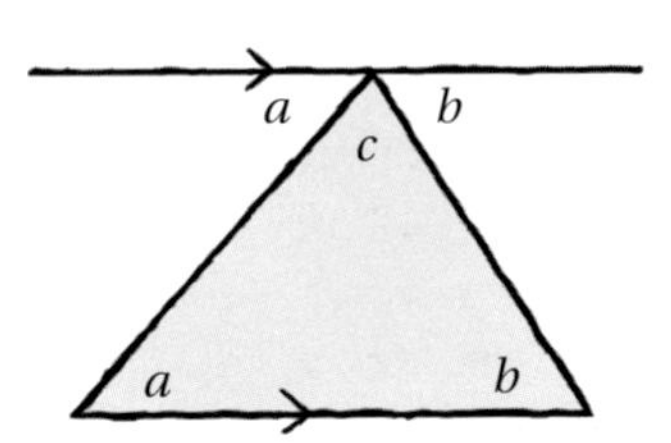

Now there are now three angles fanning out around C. What is more, these three angles must add up to 180° as they are *angles on a straight line*, in the terminology of the previous chapter.

The middle angle c is unchanged. The insight is that one of the other two is actually equal to a; this follows from the rule of *alternate angles* (or 'Z angles') that we saw in the last chapter. By exactly the same reasoning, the third angle must be equal to b. Now we have finished! The three angles at C are a, b and c, but these are angles on a straight line, and so $a + b + c = 180°$.

Triangular angular calculations

Now that we know the three angles in a triangle always add up to 180°, we can use this to perform some calculations.

To start with, if we know two angles of a triangle, we can always work out the third. For instance, if a triangle contains angles of 30° and 45°, then the final angle must the number which when added to 30° + 45° gives 180°. That is to say the final angle must be 180° − 30° − 45° = 105°.

In some cases we can do better than this. If the triangle is equilateral, then we know immediately that all three angles are equal. So it follows that each must be 180° ÷ 3, that is, 60°.

Similarly, if a triangle is isosceles, we can often work out its angles quite quickly. In isosceles triangles, two of the three angles are equal. So now we just need to be given one to work out the others. Suppose ABC is an isosceles triangle where the lengths AB and AC are the same. As usual, we'll call the three angles a, b and c. It follows that the angles at B and C must also be equal, that is to say, $b = c$. Suppose we are now told that $b = 55°$. Then we know immediately that $c = 55°$. Finally we can work out angle a. It is 180° − 55° − 55° = 70°.

Areas of triangles

We are now going to look at the area of triangles, that is, how much space there is inside them. (We will study area more generally in *Area and volume*.)

TRY THIS YOURSELF IN QUIZ 3!

There is a nice rule for calculating the area of a triangle: multiply the base of the triangle by its height, and then divide by 2. So the formula is:

$$\text{Area} = \frac{\text{base} \times \text{height}}{2}$$

Or, if we call the area A, the base b and the height h, the formula becomes:

$$A = \frac{b \times h}{2}$$

This rule is very convenient, and easy to use. But it comes with a few words of warning nevertheless!

Firstly what do these terms mean? What is the 'base' of a triangle? It is the side which runs along the bottom. All right, but here's something to think about: if we spin the triangle around, it will take up the same amount of space. So the area won't change. But the base does change! The edge which was at the bottom is no longer the base and one of the other sides takes its place.

So what's going on? In fact, any of the three sides will do as the 'base'. So this is really three formulae in one, depending on which side you pick.

Let's suppose we have chosen one side as the base. Now, what is the 'height'? This is where mistakes often get made! The height is the *vertical* distance from the base to the top corner of the triangle. That sounds reasonable, I hope, but beware:

Warning! The triangle's height may not coincide with any of its edges!

Only in right-angled triangles is the height of the triangle the length of one of its sides. In all other triangles it isn't. The rule is that the height must always be measured at right angles to the baseline: it is the vertical height from the baseline to the top corner. In fact, the height-line may not even be inside the triangle, as this picture shows!

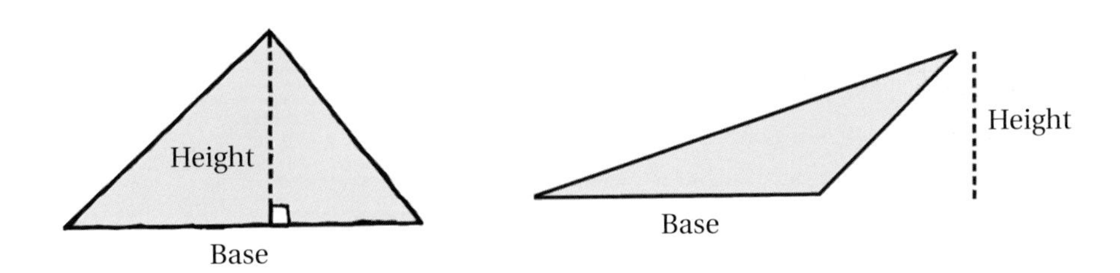

Let's have an example. Suppose I have decided to look for a house to buy, in the region between three towns: Asham, Bungleside and Cowentry (A, B, C). The question is: how large is the area I have to search? Using a map, I measure the distance from A to B as 50 miles. Taking this as the base, then the height must be the perpendicular distance from the line AB up to the third corner C. The map says that this is 60 miles. So the area of my triangle is $\frac{50 \times 60}{2} = 1500$ square miles.

CALCULATE THE AREAS TRIANGLES IN QUIZ 4.

The triangle within

The formula for the area of a triangle is undeniably useful. But why is it true? Have a look at the triangle below. There is the baseline (b), and a height-line (h). Notice that the height-line divides the shape into two smaller triangles. (As it happens, these are both right-angled triangles.)

Now, we can fit the whole triangle inside a rectangle. What is more this rectangle has the same basic dimensions as our triangle. Its height is h, and its width is b. So the area of the rectangle must be $b \times h$. (See *Area and volume* for more on this.)

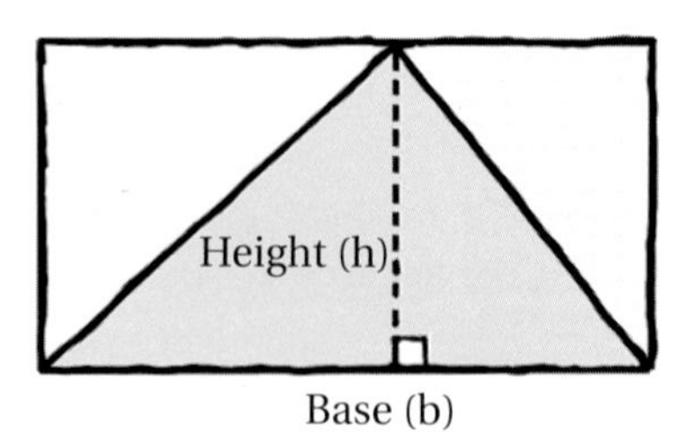

What I want to show is that the triangle takes up *exactly half* the space inside the rectangle. Why? Well, look at the two little triangles. There is an exact copy of each of them inside the rectangle. So the rectangle is divided into four triangles: two copies of each of the two small triangles, of which one is inside and one outside the original triangle. So the original triangle is exactly half of the rectangle, and its area must be $\frac{b \times h}{2}$, as expected.

Sum up *Triangles are the simplest shapes you can build with straight lines. But triangles can look very different! Luckily, there are some elegant rules for finding their angles and their areas.*

Quizzes

1 Draw a triangle, as accurately as possible and reasonably large, to fit each of these descriptions.

a An equilateral triangle **b** An isosceles right-angled triangle **c** A scalene obtuse triangle **d** An isosceles acute triangle **e** An isosceles obtuse triangle

2 Measure all the angles of each of the triangles you drew in quiz 1. Check that the three angles sum to 180°.

3 Sketch the triangles and calculate the angles.

a If a right-angled triangle also contains an angle of 30°, what is the third angle?

b ABC is a triangle where the angle at A is 120°, and the angle at B is 45°. What is the angle at C?

c ABC is an isosceles triangle, where AB = AC. The angle at B is 70°. What are the other two angles?

d ABC is an isosceles triangle, where AB = BC. The angle at B is 70°. What are the other two angles?

4 Sketch these triangles and calculate their areas.

a A triangle with base 2cm and height 1cm

b An acute triangle with base 2cm and height 2cm

c A right-angled triangle with shortest sides of 3cm and 4cm.

d A triangle in which the longest side is 5cm, and the perpendicular height from that side is 1cm

e A triangle whose base is 6cm and height is 3cm, where the height-line runs outside the triangle

Circles

- *Knowing the parts of a circle and how they are related*
- *Meeting the famous and mysterious number π*
- *Understanding the meaning of the formula πr^2*

Geometry is full of beautiful shapes, but it has often been said that the simplest and most bewitching of them all is the circle. Certainly this is among the most useful shapes for humans, and has been since the invention of the wheel many thousands of years ago.

There is much than can be said about the geometry of circles. In fact, some of the most famous formulae in science are about circles. In this chapter we will explore these mathematical marvels.

To begin at the beginning: what is a circle? The ancient Greek geometer Euclid defined it this way: first pick a spot on the ground. That will be the circle's centre. Now choose a fixed distance, say 5 feet. Then mark every point on the ground which is exactly 5 feet from the centre. The shape that emerges is a circle.

The language of circles

Circles have their own little lexicon of terms that you need to get to grips with. To begin with, the *radius* of a circle is the distance between the centre and the edge of the circle. (That was 5 feet in the example above.)

A *compass* is a useful tool for drawing circles (also known as a *pair of compasses*: that's the tool with a pin and pencil, not the one for finding North!). If you have a compass, then the distance you set between the pin and the pencil will be the radius of the resulting circle.

IF YOU HAVE A COMPASS, THEN HAVE A GO AT QUIZ 1.

Another key word is *diameter*. This is the distance across the circle, from one side to the other, passing through the middle. A little thought should confirm that a diameter can be split into two radii, meeting in the middle. This gives us our first formula for the circle: if d is the diameter of a circle and r is its radius, then the number d is r doubled. Or more concisely, $d = 2 \times r$. Omitting the multiplication sign, as usual, we have:

$$d = 2r$$

So a radius of 5cm corresponds to a diameter of 10cm, a diameter of 6 miles corresponds to a radius of 3 miles, and so on.

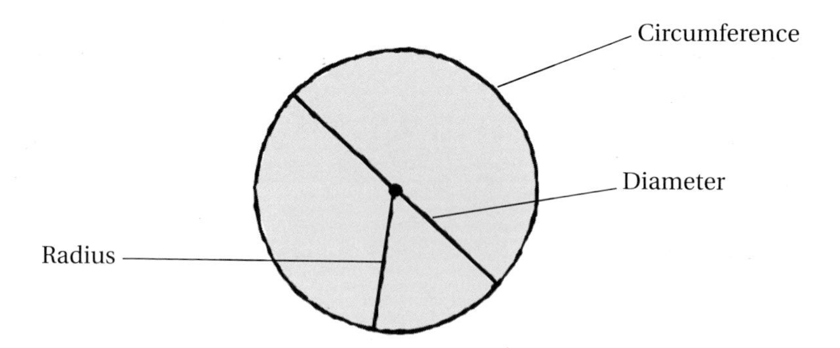

π: the legend begins

So far, so easy. But there is another significant distance which we might be interested in: the length of the circle itself, that is to say the *circumference.* If I have a circular pond in my garden, with a fountain at the centre, and a radius of 1 metre, how long is the wall around the outside of the pond?

This question, of finding the circumference of a circle when we know the radius and diameter, is a remarkably deep one, and has baffled thinkers around the world since the dawn of civilization. The ancient Babylonians, almost four thousand years ago, thought that if you multiplied the diameter by 3.125 (that is, $\frac{25}{8}$) you would get the length of the circumference. Egyptian thinkers around 1650 BC believed that the value should be $\frac{256}{81}$ (which is around 3.160). In China around AD 500 Zu Chongzhi settled on a figure of $\frac{355}{113}$, while al-Khwarizmi in ninth-century Baghdad believed it should be $\frac{3927}{1250}$.

All these geometers did at least agree on one thing: there is some number which works for all circles, however large or small. When you multiply the circle's diameter by this mystery number, you get the circumference. The only difficulty was in identifying this number exactly.

It was in 1706, at the hands of Welsh mathematician William Jones, that this elusive number finally received the name by which it is now universally known: π. Pronounced 'pi', this symbol is the Greek letter 'p' (probably chosen to stand for 'periphery').

In the centuries that followed, we have learned a great deal about π. In particular, we now understand why the geometers of old struggled with it so much. The number π is an example of what is today known as an *irrational number*. This means that it can never be written exactly as a fraction of two whole numbers. ('Irrational' here means 'not a *ratio*'; it has nothing to do with rationality in the sense of being sensible, intelligent or logical.)

This immediately means that all the old values attributed to π must be wrong because they were fractions (although some were excellent estimates, and were near enough for practical purposes). What about a decimal? We can certainly start writing out the value of π:

$$3.14159265358979323846264338327 95\ldots$$

The interesting thing is that this sequence of numbers will continue forever, never ending, and never getting caught in a repetitive loop (unlike the recurring decimals we came across when converting fractions to decimals). It simply keeps going, ever unpredictable. Hence we can never write down the value exactly, except, of course, under its name: 'π'. (The drive to calculate ever more digits of π has now reached the trillion mark, and, like π itself, is set to continue indefinitely.)

All the way around: the circumference

Interesting though the history is, the mystery of π is now largely solved. In particular, it is now easy to calculate the circumference of a circle from its diameter: you simply multiply by π. So we get the next formula for a circle: $c = \pi \times d$, where c is the circumference and d is the diameter. Equivalently, because the diameter is twice the radius (r), we might say $c = 2 \times r \times \pi$, or omitting the multiplication signs, and reordering:

$$c = 2\pi r$$

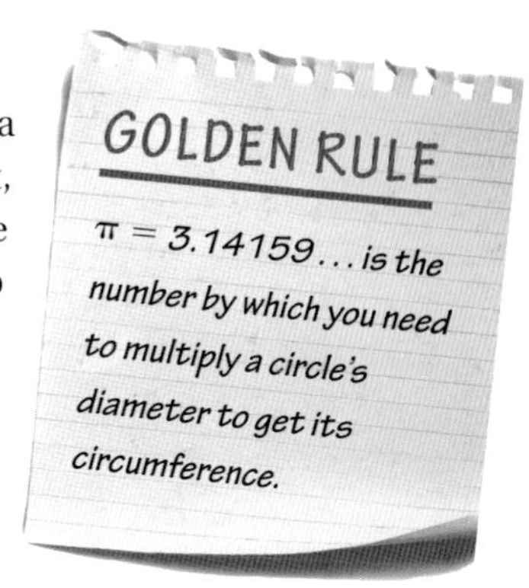

For most of us, the correct tool for using this formula is the [π] button on a calculator. Though not exact, this will give as good an approximation of π as we will ever need. (In some calculators you may have to press SHIFT or 2nd FN and then another button to access the π function.) So, to return to the example of the pond above, if my circular pond has a radius of 1 metre, then its circumference is $2 \times \pi \times 1$. Typing in [2] [×] [π] [=] to my calculator produces an answer of 6.28 metres (when rounded to two decimal places; see *Decimals*).

Turning this round, if we know that a circle has a circumference of 10 metres, how can we calculate its radius? In other words, we have to find the value of r so that $2 \times \pi \times r = 10$. To solve this, we just need to divide 10 by $2 \times \pi$ (see *Equations* if you have forgotten why this works). On my calculator I do this by typing [1] [0] [÷] [(] [2] [×] [π] [)] [=] but other calculators may work differently. This brings up an answer of 1.59 metres, to two decimal places. (The brackets are needed to make sure that I divide 10 by twice π, rather than dividing by 2 and then multiplying the answer by π.)

TRY SOME CALCULATIONS YOURSELF IN QUIZ 2.

So now you know how the radius, diameter and circumference of a circle are all related.

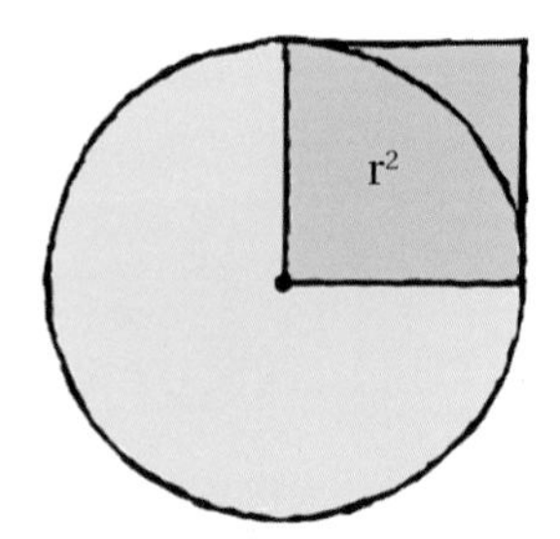

Area

We might also want to know the *area* of a circle, that is, the amount of space it occupies. For some shapes, such as a square, finding the area is easy: if the square's side is 3cm long, then its area is $3 \times 3 = 9\text{cm}^2$. Once again, though, the circle is less straightforward, and again the number π takes centre stage.

If a circle has radius $r = 3\text{cm}$, then what is its area? If we build a little square on the radius of the circle, we know that its area is $r \times r$, or r^2 for short. That comes out as $3 \times 3 = 9\text{cm}^2$ in the above example.

So, the question is, how many times does this little square fit inside the circle? The answer - for any circle, large or small - is again π. So the area of the circle is given by $\pi \times r \times r$. Calling the area A, this gives us one of the most famous of all formulae:

$$A = \pi r^2$$

In the example above, where the circle has radius 3cm, the area is $\pi \times 9 = 28.27\text{cm}^2$, to two decimal places.

CALCULATE THE AREA OF SOME CIRCLES IN QUIZ 3.

Going backwards

We might want to work backwards. If we know the area of a circle, how can we work out its dimensions? Suppose a circle has an area of 4cm^2. If we call its radius r, it must be that $\pi r^2 = 4$. This is now an equation, where our job is to find out r. (See *Equations* for more discussion of this.) We begin by dividing both sides by π, to get $r^2 = \frac{4}{\pi}$. We could now put this into the calculator, but let's first see how to finish the calculation off. We now know what r^2 is. To find out r from this information, we need to take the square root (see *Roots and logs*) of the number we have just found. So the exact answer is $r = \sqrt{\frac{4}{\pi}}$

There are various ways to get a value for this, depending on your calculator. One way would be [√] [(] [4] [÷] [π] [)] [=]. Again the brackets are crucial. On an older calculator, the best approach might be [4] [÷] [π] [=] followed by [√] [=] (or maybe [√] [Ans] [=], where [Ans] is the button which recalls the answer to the previous calculation). Whatever method you use, the answer should come out as 1.13cm, to two decimal places.

TRY QUIZ 4, AND THE MORE CHALLENGING QUIZ 5.

Sum up *Circles are among the most beautiful and useful shapes. They are also associated with some of the most beautiful and useful of all mathematics!*

Quizzes

1 Use your compass to draw circles with these sizes. Then, for each circle, estimate the length of the circumference (that is, the distance around the outside).

a A circle with a radius of 2cm

b A circle with a diameter of 2cm

c A circle with a radius of 4cm

d A circle with a diameter of 4cm

e A circle with a radius of 1cm

2 Calculate these lengths. Give your answers to two decimal places.

a The circumference of a circle with diameter 5cm

b The circumference of a circle with radius 5cm

c The radius of a circle with diameter 7 miles

d The diameter of a circle with circumference 50km

e The radius of a circle with circumference 4.4mm

3 Calculate these areas. Give your answers to two decimal places.

a The area of a circle with radius 5cm

b The area of a circle with diameter 5cm

c The area of a circle with radius 2.13mm

d The area of a circle with diameter 2.13mm

e The area of a circle with circumference 2.13mm

4 Calculate these.

a What is the radius of a circle with area 5 square miles?

b What is the diameter of a circle with area $13mm^2$?

c What is the circumference of a circle with area $5.3km^2$?

d A circle fits inside a square exactly (touching but not crossing all four sides). If the area of the circle is $8cm^2$, what is the area of the square?

e Overnight, a pattern of flattened crops appears in a farmer's field, consisting of four non-overlapping circles, all the same size. If the total area of flattened crops is $700m^2$, what is the radius of each circle?

5 Concentric circles are circles with the same centre.

a A picture consists of two concentric circles with radii 3cm and 4cm respectively. What is the area of each?

b The circular strip between the two circles in part a is painted red. What is the area of the red stripe?

c A circle fits inside a square exactly. The width of the square is 5cm. What are the areas of the square and the circle?

d In part c what is the total area of the parts of the square outside the circle?

e Inside a circle of radius 6cm is a triangle, whose base is a diameter of the circle, and whose height is a radius. What is the total area of the parts of the circle outside the triangle?

Area and volume

- *Understanding how length, area and volume are related*
- *Knowing how to calculate areas and volumes for different shapes*
- *Getting to grips with the units for area and volume*

***We have many units for measuring distance: miles, millimetres, yards, parsecs, furlongs, inches, light-years, metres, feet, . . . and that ignores archaic measurements such as rods, leagues, paces and perches. In this book we will mostly stick with metres, and related units such as millimetres, centimetres, and kilometres (see* The power of 10*). Any measure of distance automatically produces a related measurement of* area.**

Distance is one-dimensional: it measures a length along a line. Area, on the other hand, is two-dimensional. It measures the size of a surface.

What is the relationship between length and area? If a metre (1m for short) is our basic unit of distance, then the corresponding unit of area is the square metre (or metre squared), $1m^2$. By definition, this is the area of a square which is 1m long and 1m wide.

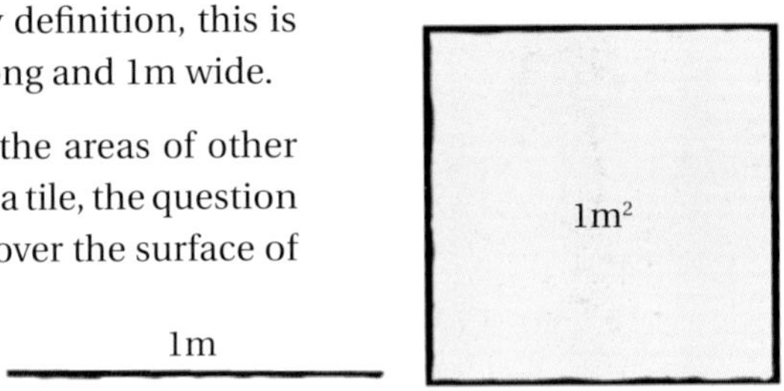

We can then use this to quantify the areas of other things. Thinking of a $1m^2$ square as a tile, the question is: how many tiles are needed to cover the surface of other shapes and objects?

1m

$1m^2$

Of course, we don't have to use whole tiles, they can be cut up as necessary. A triangle which is 1m wide and 1m high occupies exactly half of one of these square tiles, and so has an area of $\frac{1}{2}$ m^2.

For smaller shapes, we might want to use cm^2 instead, meaning tiles 1cm by 1cm, while for measuring the areas of whole countries, tiles of $1km^2$ are more appropriate. When lengths are quoted in other units, such as inches or miles, it makes sense to use the corresponding units for area: square inches or square miles. It is good to be flexible about such things. (Beware, though: the relationship between cm^2, m^2 and km^2 is not completely obvious! More on this later.)

The road outside my house has an area of around $500m^2$, meaning that 500 tiles of size $1m^2$ are needed. A (large) adult human being might have a surface area of around $2m^2$, which tells us how much skin they have.

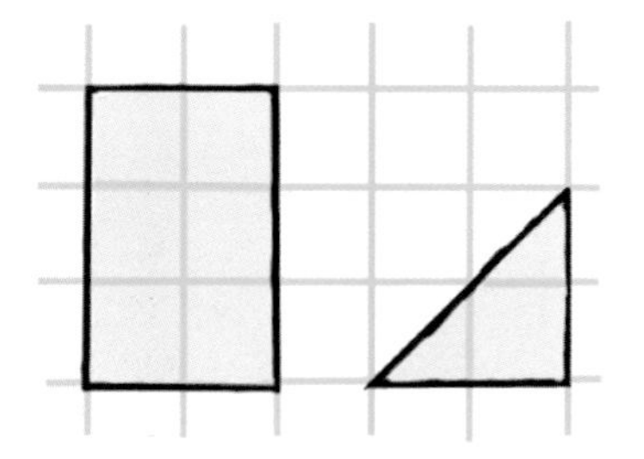

The first way to calculate (or estimate) the area of a shape is just by counting the number of tiles needed to cover it. For instance the rectangle on the left is covered by six tiles, while the triangle is covered by one whole tile and two half tiles, making a total of two tiles.

Calculating, not counting

Some shapes have areas which can be calculated exactly, quickly and easily, without the laborious task of counting tiles. My road is 100m long and 5m wide. What this means is that 100 tiles of 1m^2 would fit along its length, and 5 tiles would fit across its width. So the road would be completely covered by 5 rows of 100 tiles, which is why its area is $100 \times 5 = 500\text{m}^2$. Notice that I can calculate this from just two simple measurements, without having actually to work with tiles.

This is an example of a broader phenomenon. To calculate the area of a *rectangle*, you can multiply its two distances together: area = length × width. A rectangle which is 8cm long and 3cm wide has area $8\text{cm} \times 3\text{cm} = 24\text{cm}^2$.

A square is a special case of this, because a square is nothing more than a rectangle whose length and width are the same. So a square which is 5cm wide has an area of $5\text{cm} \times 5\text{cm} = 25\text{cm}^2$.

CALCULATE THE AREAS OF SOME RECTANGLES IN QUIZ 1

Squared units versus units squared

How many cm^2 are there in 1m^2? The obvious answer might be 100, since, as most people know, there are 100cm in a metre. But the obvious answer can often be wrong! In fact, a square which is 1m × 1m is 100cm × 100cm, so its area in cm^2 is $10{,}000\text{cm}^2$.

Another, similar, source of confusion is the difference between a 'hundred square kilometres' (meaning 100km^2) and a 'square of a hundred kilometres' (meaning 100km × 100km). How wide is a square whose area is 100km^2? Is it 100km? No, the correct answer is 10km, because $10 \times 10 = 100$. A square which is 100km long would have area of $100\text{km} \times 100\text{km} = 10{,}000\text{km}^2$.

Now, how many m^2 are there in 1km^2?

Beyond rectangles

The areas of rectangles are easy to calculate if we know their lengths and widths. But other shapes are trickier. Some have their own rules for calculating their areas. To calculate the area of a triangle, for example, we need to multiply its height by its base and then divide by 2 (see *Triangles*). This sounds easy enough, but remember the warning: the height of a triangle may not be the length of any of its sides! The height and the base must always be at right angles to each other.

The area of a circle is given by the famous formula πr^2 (see *Circles*). Here r is the circle's radius and π is the number 3.14159. . . So if a clock face has radius 5cm, then its area is $\pi \times 25\text{cm}^2$, which is 78.5cm^2, to one decimal place.

We use rules such as this to work out the areas of some other related shapes. To get the area of a semicircle for example, we can calculate the area of a circle, and then divide by 2. To calculate the area of a ring, we can subtract the area of the inner circle from the area of the outer circle.

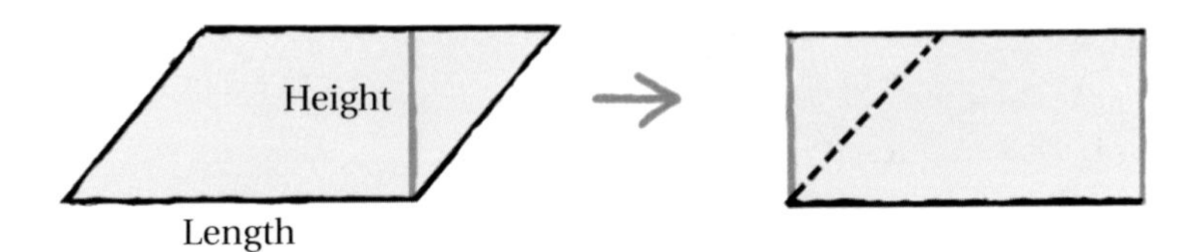

Beyond squares and rectangles, other *quadrilaterals* (four-sided shapes) have their own rules for area. (See *Polygons and solids*.) The area of a *parallelogram* is the same as the area of a rectangle: the length times the height. Be careful though because, as with a triangle, the height of a parallelogram is not the length of any of its sides, but the perpendicular distance between opposite sides. To see why this rule should hold true, notice that if we cut the triangular end off a parallelogram, and fit it onto the other end, the two pieces together form a rectangle.

Volume: the science of space

Just as area measures a region in two dimensions, so *volume* measures how much space objects take up in 3-dimensional space. This is useful, since 3-dimensional space is where we live. So volume is the right way to measure how big a physical object really is.

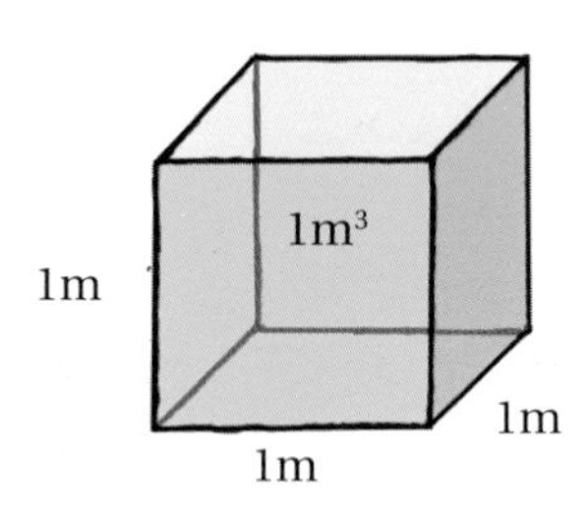

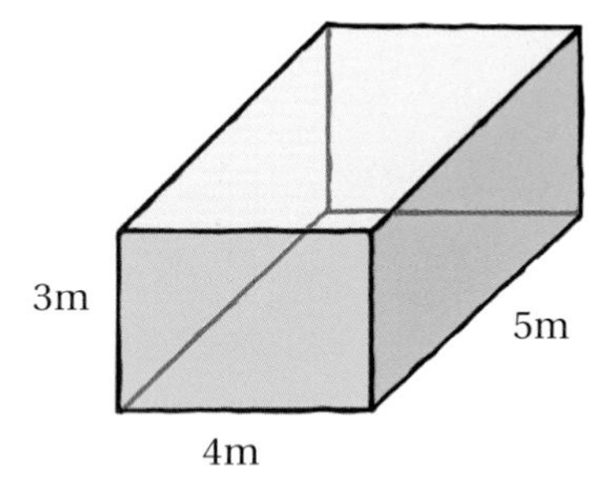

Just as a unit of length such as a metre gives us a way to measure area (square metres), it also provides us with a unit of volume, the *cubic metre* or *metre cubed*: $1m^3$. One cubic metre ($1m^3$) is the volume of a cubic box which is 1m tall, 1m wide and 1m long.

A *cuboid* is the 3-dimensional version of a rectangle, where six rectangular faces meet at right angles. We can calculate the volume of such a shape by multiplying its three dimensions. So if my room is 3m high, 4m wide and 5m long, then its volume is $3m \times 4m \times 5m = 60m^3$.

Volume is useful for measuring quantities of liquid (or gas). We often use other units of volume for this. We might talk about the capacity of a bottle in terms of *litres*, for example. There is a relationship between litres and metres. One cubic centimetre ($1cm^3$) is the volume of a $1cm \times 1cm \times 1cm$ cube, and one litre is $1000cm^3$.

So how many litres are there in $1m^3$?

The volumes of solid objects

Just as with area, many different shapes come with their own rules for calculating volume. To use these rules, we need to be able to substitute numbers into formulae. So have a look at *Algebra*, if you need a refresher on how to do that.

A *sphere*, for example, has a volume of $\frac{4}{3}\pi r^3$, where r is the radius of the sphere (meaning the distance from the centre to the edge). So if an inflatable globe has a radius of 5cm, then the amount of air it contains is $\frac{4}{3} \times \pi \times 5^3 \text{ cm}^3 = 523.6\text{cm}^3$ (to one decimal place).

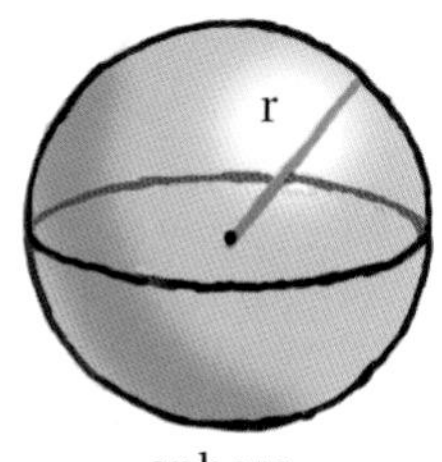

sphere

A *cylinder*, on the other hand has a volume given by multiplying the area of its circular end by its height. The area of the circular end is πr^2, where r is the radius. If h is the length of the cylinder, then putting this all together, we get a volume of $\pi r^2 h$.

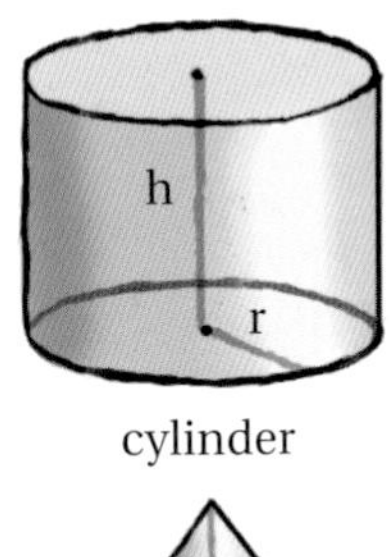

cylinder

Similarly, a *cone* (with a circular base) has a volume of $\frac{1}{3}\pi r^2 h$, where r is the radius of the circle at the bottom and h is the cone's height. (The cone occupies exactly a third of the volume of a cylinder with the same radius and height.)

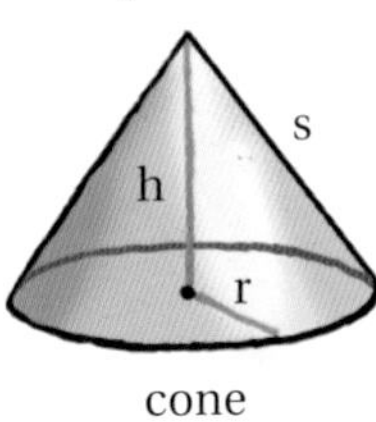

cone

So, if a wafer cone is to be filled with ice cream, and the cone is 10cm long, with a radius at its widest end of 3cm, then the amount of ice cream the cone can contain is $\frac{1}{3} \times \pi \times 3^2 \times 10 = 94.2\text{cm}^3$, to one decimal place. (Of course this ignores any ice-cream sticking out beyond the end of the cone!)

We might also need the length s, which runs diagonally from the base to the point. (If necessary s can be calculated from r and h using Pythagoras' theorem, covered in a later chapter.) It turns out that the area of the curved part of the cone is πrs.

It might be useful to put this information into a grid:

	SPHERE	CYLINDER	CONE
Surface area	$4\pi r^2$	Curved surface: $2\pi rh$ Flat surfaces: $2\pi r^2$	Curved surface: πrs Flat surface: πr^2
Volume	$\frac{4\pi r^3}{3}$	$\pi r^2 h$	$\frac{1}{3}\pi r^2 h$

Notice that, in each case, the number of lengths multiplied together to get the surface area is *two* (whether that be r^2, $r \times h$ or $r \times s$), along with some fixed constant number (π, 2π, etc.). For the cone and the cylinder two such terms need to be added together, to deal with the curved parts and the flat ends.

USE THE FORMULAE IN THE GRID IN QUIZZES 2 AND 3.

For the volume, the number of lengths to be multiplied together is *three*, whether that is r^3 or $r^2 \times h$, again multiplied by some constants. So, even among these sophisticated formulae, our golden rule holds good!

Eureka!

What is your volume? Unless you are perfectly spherical, cylindrical or conical, it is unlikely that there is a neat formula to calculate the answer. Human beings have very irregular and uneven shapes. But this is not just true of us; most objects in the real world differ from the idealized figures of geometers' dreams. So how can we calculate volumes of irregular shapes?

There is a simple method, discovered by Archimedes. According to legend, while he was having a bath, he noticed that, as he got into the tub, the water level rose. Many people must have noticed the same thing. But Archimedes asked himself one simple question: by *how much* did the water rise?

It was then that Archimedes uttered his famous cry 'Eureka!'. He realized that the volume of bathwater displaced by his body, must be exactly equal to his body's volume.

To transfer the action out of the bath and into the lab, suppose we have a measuring cup, with some water in it. The scale tells us that the volume of the water in the cup is 100cm^3. Now suppose we have some object, whose volume we want to know. It doesn't matter how strangely shaped it is, or what it is made of (so long as it isn't absorbent). If it fits in the cup, we can measure it. When we put it in the water, the water level rises. Let's say it now reads 123cm^3. Then we immediately know that the volume of the object is 23cm^3.

With this simple piece of equipment, we can now calculate volumes using only *subtraction*. (The only thing to watch out for is that the object is fully submerged. If part of it sticks out of the water, that part will be missed out of the calculation.) Why not try it?

Sum up *Areas and volumes both represent the size of objects, but in different dimensions. With a bit of practice, calculating them is not hard!*

Quizzes

1 Calculate the area of each of the objects in parts a–d.

a A fencing strip 14m long and 2m wide

b A football pitch 100m long and 50m wide

c A computer screen 30cm wide and 20cm high

d A square photograph 4cm wide.

e There are 12 *inches* in a *foot*. How many *square inches* are there in a *square foot*?

2 Work out these volumes.

a A fish tank is a cube, 50cm wide. How much water can it hold?

b A block of flats is a cuboid 40m high, 20m wide and 15m deep. What is its volume?

c An award trophy is a solid gold globe with radius 4cm. How much gold is needed to make it?

d A flotation device is a plastic cylinder that is 1 metre long and has a circular end with a radius of 10cm. How much air does it contain?

e A wizard's hat is a cone, 30cm high, with a radius of 10cm at the circular base. What is its volume?

3 What are the surface areas of the shapes in quiz 2? (For part e you will need Pythagoras theorem. Come back to this after the chapter on Pythagoras, if you are not yet familiar with that.)

Polygons and solids

- *Knowing the different possible shapes with four sides*
- *Understanding how the geometry of regular polygons works*
- *Recognizing the Platonic solids*

Geometry is the study of shapes. But what shapes are there? We have already met the circle and the triangle. In this chapter we take a whistle-stop tour around the wider world of geometry, to see some of the other shapes that are out there. It's going to be fast and furious, so hang on to your hats! Our first stop is the quadrilateral zoo.

The quadrilateral zoo

As we saw earlier, the humble triangle comes in all sorts of different forms: obtuse, acute, right-angled, isosceles, and so on. *Quadrilaterals*, or four-sided shapes, exhibit an even greater variety. So, to begin this chapter, we will have a nose around the quadrilateral zoo.

Its most famous inhabitant is the familiar, dependable *square*. This shape is as symmetric as possible: the four sides all have the same length, and the four angles are all right angles. Although the square itself is a very familiar shape, a slight alteration can change its appearance. When it's standing on one of its corners instead of flat on its edge, it often gets called by a different name: a *diamond*. But of course the two are the same shape!

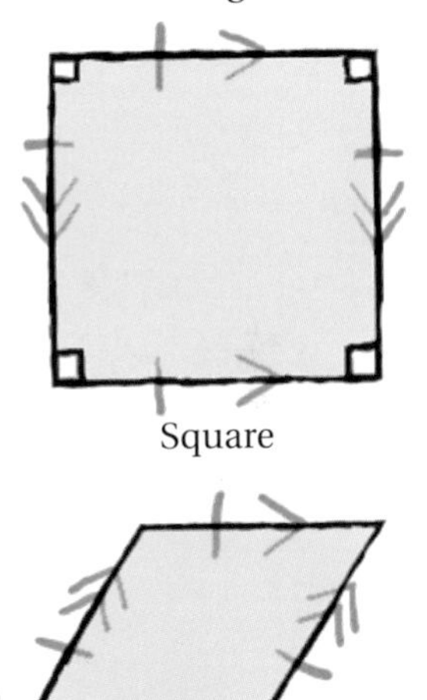
Square

Rhombus

The square also serves to illustrate some of the little symbols geometers use to communicate facts about shapes. Parallel lines (as we saw in *Angles*) are marked by matching arrows. In the square, both pairs of opposite sides are parallel. We also use matching notches to indicate lines that are the same length. (Of course, in a square, all four sides are the same length.) Finally, there is the little mark we use to indicate a right angle. In a square, all four angles are right angles.

Now let's visit one of the square's closest cousins, the *rhombus*. Like a square, a rhombus has all its four sides the same length. But, unlike the square, its angles need not be all the same, and need not be right angles. So a rhombus looks like a square which has been squeezed.

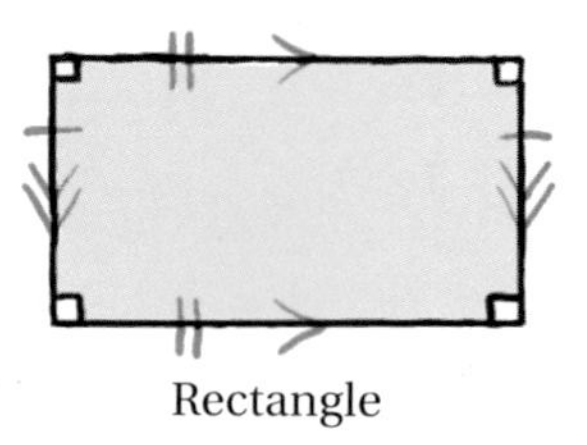

Rectangle

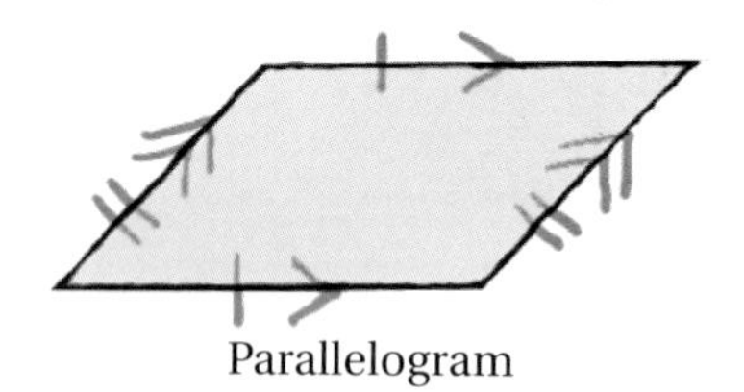

Parallelogram

On the other side of the family, the square's nearest relation is the *rectangle*. Like a square, a rectangle has four right angles inside. The difference is that the lengths of its edges are not all the same. They come in two pairs, with opposite edges of equal length. So a rectangle has two long edges opposite each other, and two shorter edges opposite each other.

A *parallelogram* has the same relationship to a rectangle as a rhombus does to a square. It looks like a rectangle which been squashed, so that its edges no longer meet at right angles. The name 'parallelogram' comes from the fact that sides which are opposite each other are always parallel (as well as being the same length).

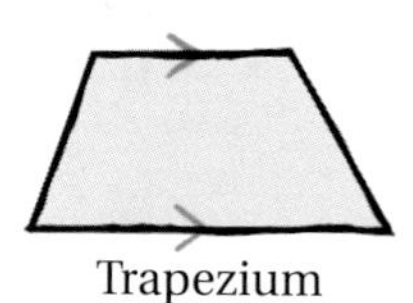

Trapezium

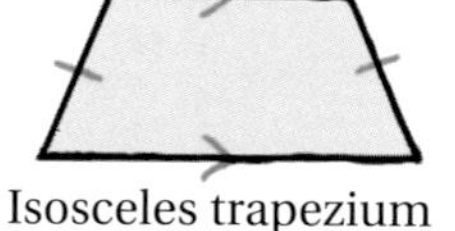

Isosceles trapezium

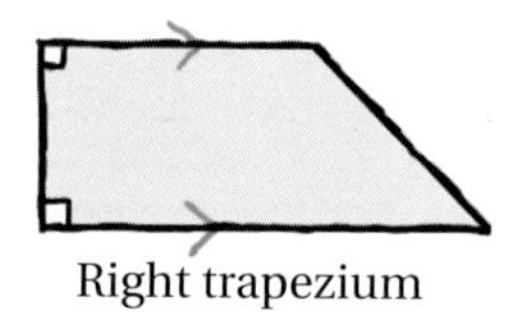

Right trapezium

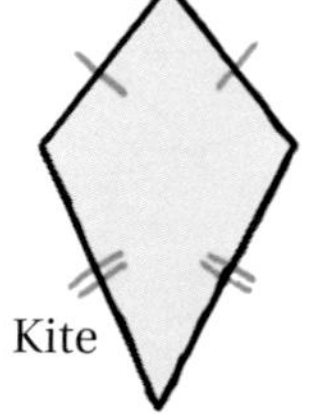

Kite

A *trapezium* (or *trapezoid*) is another shape which is built from a pair of parallel lines. Unlike in a parallelogram, the remaining two sides do not need to be parallel, or the same length. There are several subspecies of trapezium: a *right trapezium* is one which contains two right angles, so that one of its ends looks like the end of a rectangle. An *isosceles trapezium* is one which is symmetric: the two non-parallel side-lines are the same length, and meet the parallel lines at the same angles. (An isosceles trapezium looks like an isosceles triangle with the top cut off.)

A *kite* is a shape which is built from two pairs of lines of equal length: two short lines and two long lines. In that sense it is like a parallelogram. The difference is that, in a parallelogram, the identical lines are opposite each other and are parallel. In a kite, the identical lines are neighbours and are not parallel.

Some of the inhabitants of the quadrilateral zoo are more exotic creatures, what mathematicians call *reflex quadrilaterals*. Remember from *Angles* that a *reflex angle* is one that is bigger than 180°. These figures might look like a larger shape which has had a bite taken out. The most symmetric reflex quadrilateral is an *arrowhead* or *chevron*. The definition of this is actually the same as for the kite; it is just that the two shorter sides meet at a reflex angle. (They go 'in' instead of 'out'.)

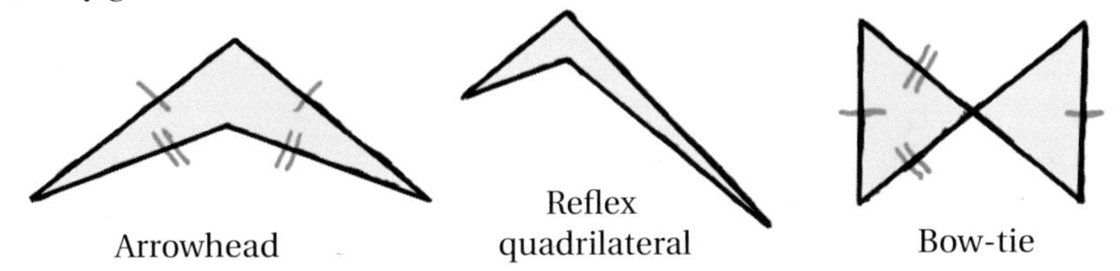

The strangest creatures in the zoo are the *self-intersecting quadrilaterals*. Not everyone agrees that they should even be allowed in. These are shapes in which two of the edges crash through each other. The most symmetric of these are the *bow-ties*. A bow-tie looks like a rectangle which has been assembled badly: one pair of edges, instead of running parallel, crosses over.

SEE IF YOU CAN IDENTIFY THE DIFFERENT SPECIES IN QUIZ 1.

I hope you have enjoyed your visit to the quadrilateral zoo!

Shapes with more sides: polygons

We have now investigated three-sided shapes (triangles) and four-sided shapes (quadrilaterals). These fit into the broader category of *polygons*, the general name for flat shapes built from straight lines.

As we look at polygons with more and more sides, the variety is only going to increase. So, from now on, we will focus on only the most symmetric shapes.

Of all the different types of triangle discussed in the chapter on triangles, the most symmetric are the triangles where all the sides are the same length, and all the angles are equal. These are the *equilateral* triangles.

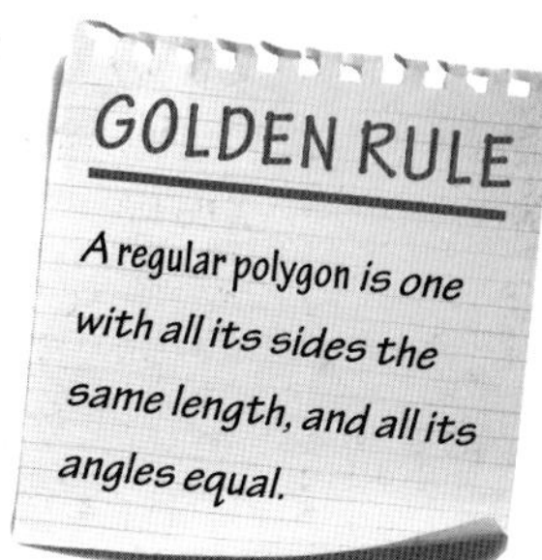

Similarly, of all the quadrilaterals discussed above, the most symmetric is the *square*. Again, all its sides are the same length, and all its angles are equal.

We can continue this pattern with this chapter's golden rule.

That's fine, but after the equilateral triangle, and the square, what other examples are there? There is the *regular pentagon* with five sides, the *regular hexagon* with six sides, the *regular heptagon* with seven sides, and so on. (The words here have their origin in the ancient Greek numbers.) At each level there are also countless *irregular* possibilities. Just as the square is not the only quadrilateral, so there are endless irregular pentagons and hexagons, and so on.

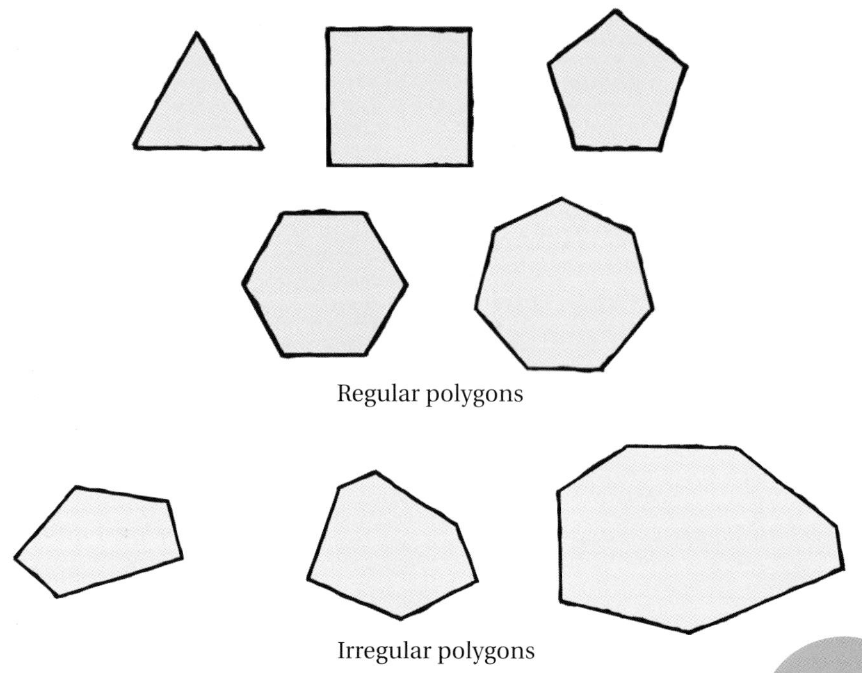

Regular polygons

Irregular polygons

Regular polygons

DRAW SOME DIFFERENT SHAPES IN QUIZ 2.

An equilateral triangle has angles of 60°, a square has angles of 90°. What is the pattern here? What is the angle inside a regular pentagon, for example?

There is a nice way to answer this question. Suppose you are walking along the outside of a regular pentagon. You turn five corners, each the same angle. Then, at the end, you are back facing the same way. So you must have turned exactly 360° altogether. Each corner then, must be $\frac{360}{5} = 72°$. So is this the answer? Not quite!

The corner you turn when walking around the pentagon is *not* the shape's internal angle. It is the amount by which the next edge along deviates from the line going straight on (see the diagram). This is sometimes called the shape's *external angle*, and it is this that we have calculated to be 72°.

To find the internal angle, we need to know the relationship between that and the external angle. But the diagram shows it: they are angles on a straight line. That means that they must add up to 180°. So if the external angle is 72°, then the internal angle must be 180° − 72° = 108°.

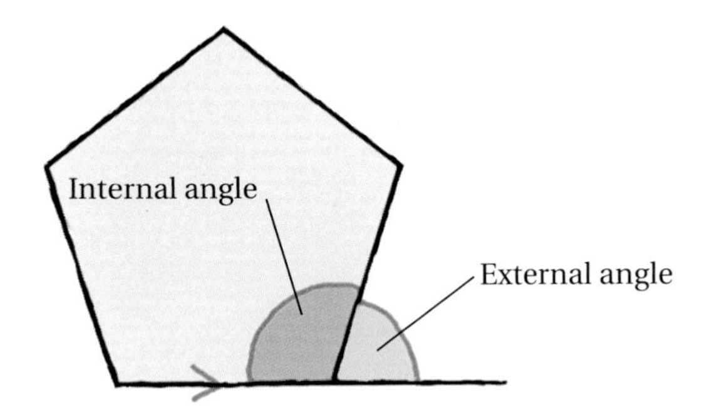

TRY IT FOR OTHER REGULAR POLYGONS IN QUIZ 3.

The method we have used to calculate the angle inside a regular polygon involves two steps:

- Divide 360° by the number of sides of the polygon to find the external angle.
- Subtract the external angle from 180° to find the internal angle.

Into the 3D world: solids

For the final stop on this geometrical mystery tour, we will take a bold step out of the flat land of two dimensions into the world of three dimensions. Here the counterparts of polygons are known as *polyhedra*, or simply *solids*. The most familiar example is the *cube*, the big brother of the two-dimensional square.

Polygons were two-dimensional shapes built from straight lines, which met at corners. A *solid* is a three-dimensional shape built from flat faces which meet at straight edges and corners. Other examples of solids are *cuboids*, which are the three-dimensional equivalents of rectangles, *pyramids*, and a host of others.

Just as we concentrated on regular polygons above, now we shall focus our investigation on *regular* solids. This is the setting for one of the greatest and most ancient of all mathematical theorems: the classification of the Platonic solids.

The Platonic solids

Regular polygons were fairly easy to understand: we began with equilateral triangles (three sides), then looked at the square (four sides), the regular pentagon (five sides), regular hexagon (six sides), and so on. For every number after that, there is a corresponding regular polygon, and we have a very good idea of what it will look like. There are no surprises in store!

The rules of the solid world are less clear, but the golden rule for polygons does have a counterpart here:

> *A regular solid (or regular polyhedron) is one in which all the faces are identical regular polygons, all the edges are the same length and all the angles are the same.*

That's an elegant definition. But the big question is: apart from the cube, what other examples are there?

This is a question which preoccupied the geometers of ancient Greece, including the philosopher Plato. As he was able to prove, while there are infinitely many regular polygons, the collection of regular solids is a much smaller family. Let's work through the reasoning that led Plato to that conclusion.

One of the simplest regular solids, aside from the cube, is the *tetrahedron*. This is a pyramid with a triangular base. Just as a cube is built from six squares, a tetrahedron is build from four equilateral triangles (in fact its name means 'four sides').

What are the broader rules here? The faces of any regular solid must be regular polygons; that much is clear. So having built one from triangles, and one from squares, a natural question is: can we build one from pentagons? The answer is yes! Meet the dodecahedron (meaning 'twelve sides'), with its 12 pentagonal sides, which come together in threes at every corner.

But when we try to carry on with this line of thought, we are in for a surprise. There is no regular solid that can be built from hexagons.

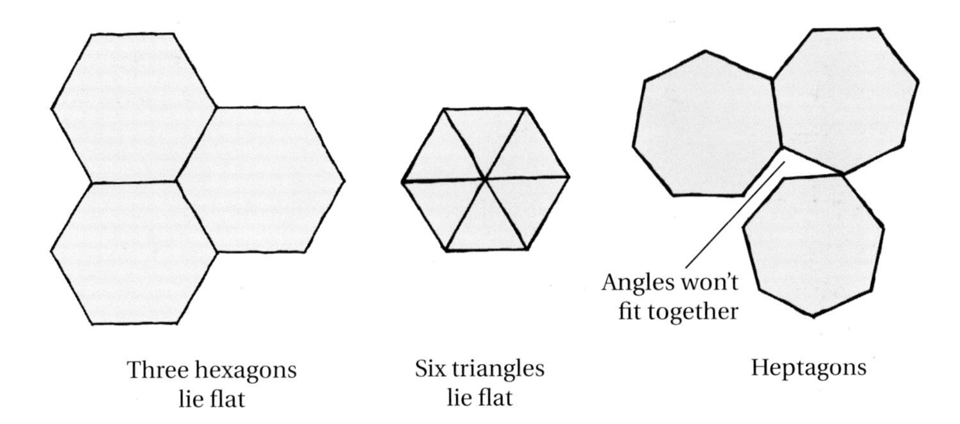

Three hexagons lie flat

Six triangles lie flat

Heptagons

Let's see why. At every corner of a solid shape, some faces meet. There must be at least three of them: two cannot be enough. But if we take three hexagons, they mesh together perfectly on a flat plane. So there is no way of fitting them together to make a shape with any depth to it. There is no way to fit together any more than three either. (If you doubt it, try it!)

A similar argument shows that heptagons, octagons, and so on, can never be fitted together to make a regular solid: once you have put the first two together, there is never enough space for a third.

So the options for building regular solids are limited to triangles, squares and pentagons. A cube is the only solid which can be built from squares, since the only option is for three squares to meet at each corner (as four squares connected at a corner would lie flat). Similarly the dodecahedron is the only possibility for pentagons. The tetrahedron is constructed from triangles, but maybe there are other solids which can also be built from triangles?

Yes! In a tetrahedron, three triangular faces meet at each corner. But in an *octahedron*, four do. The octahedron has eight faces in total, and looks like two square-based pyramids glued together.

There is another shape in which five triangular faces meet at each corner: this is the *icosahedron*, which has 20 faces in total.

When we fit six triangles together, though, they lie flat. So we have now exhausted every possibility.

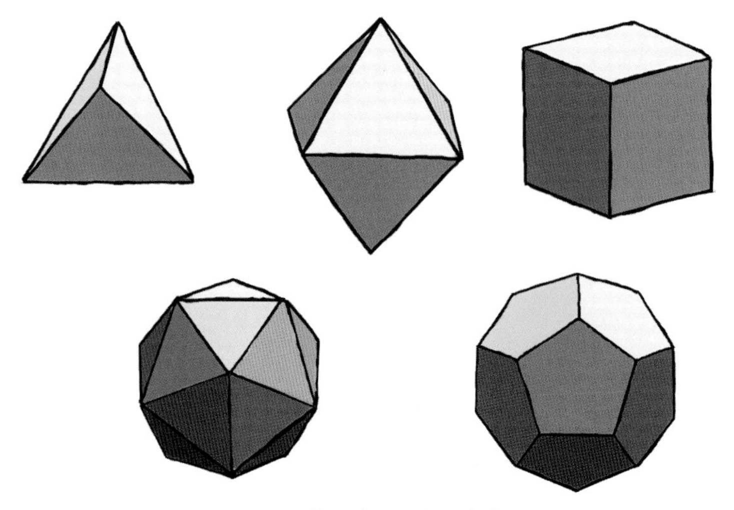

The Platonic solids

At the end of all that we have found a family of five regular solids: the tetrahedron, octahedron, cube, dodecahedron, and icosahedron. These are known as the Platonic solids.

The beauty of this theorem is not only that we have found some very pretty shapes, but that we have completed the catalogue. This is a full list; besides these five shapes there are no other regular solids!

EXPLORE THE PLATONIC SOLIDS FURTHER IN QUIZ 4

Sum up *Whether in two dimensions or in three, and whether studying regular solids or irregular quadrilaterals, geometry is jam-packed with fascinating facts about beautiful shapes!*

Quizzes

1 Use a ruler to draw these quadrilaterals.

a A rectangle with sides of 2cm and 3cm
b A rhombus with sides of 1.5cm
c A parallelogram with sides of 4cm and 2cm
d A right-trapezium with parallel sides of 2cm and 5cm
e A kite with sides of 3cm and 5cm

2 Draw a picture of each of these shapes.

a an irregular pentagon (five sides)
b a right trapezium
c a regular hexagon (six sides)
d an irregular reflex quadrilateral
e an irregular ikosihenagon (21 sides)

3 Find the internal angle of each of these regular polygons.

a a regular hexagon (six sides)
b a regular heptagon (seven sides)
c a regular octagon (eight sides)
d a regular decagon (ten sides)
e a regular hexekontakaitriakosiagon (360 sides)

4 Work out the number of corners, edges and faces of each of the Platonic solids. (Hint: in a tetrahedron, each face has three edges. But each of these edges is shared between two faces. So the total number of edges is 3 times the number of faces, divided by 2.)

Pythagoras' theorem

- *Understanding what Pythagoras' theorem means*
- *Knowing how the theorem can be used to calculate lengths*
- *Recognizing how right-angled triangles are used in the wider world*

Living in Greece in the sixth century BC, Pythagoras was one of the most famous mathematicians of the ancient world. The theorem which bears his name is one of the first great facts of geometry. Actually it is a matter of debate whether this theorem can be attributed to him personally rather than the group of thinkers whom he inspired, the Pythagoreans. The theorem may even have been known to earlier geometers in India or Egypt.

Leaving the history aside, whoever first proved it, Pythagoras' theorem is about triangles, those friendly three-sided creatures that we met in an earlier chapter. Specifically, it is about *right-angled triangles*. These are in many ways the most interesting triangles.

To start with, what is a right-angled triangle? Very simply, it is a triangle which contains a right angle. That is to say, one of the three angles in the triangle is equal to 90°. We can also see a right-angled triangle as half of a rectangle which has been sliced in half diagonally. This perspective will be useful later on.

Pythagoras' theorem

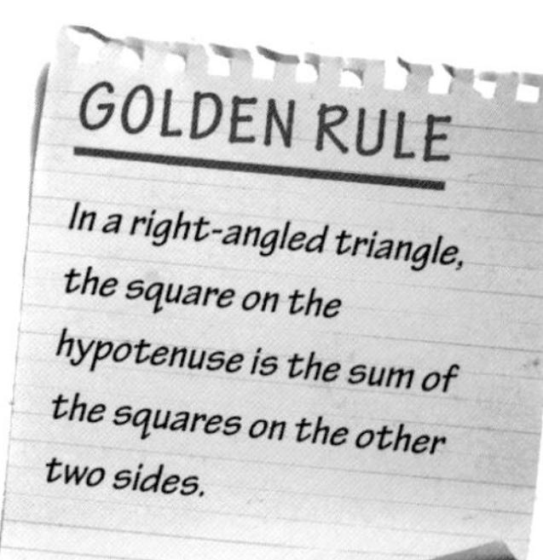

The main fact we discovered in *Triangles* was about the *angles* in a triangle. Pythagoras' theorem describes a relationship between the *lengths* of the three sides of a right-angled triangle (not any other kind of triangle). Let's call the three lengths a, b and c, where c represents the longest side. In a right-angled triangle, the longest side is always the one opposite the right angle. It even has a fancy name: the *hypotenuse*.

Pythagoras' theorem describes a relationship between the *squares* of the three sides, that is, the numbers you get by multiplying each length by itself: $a \times a$, $b \times b$ and $c \times c$, or a^2, b^2 and c^2, for short. (See *Powers* for more discussion of squares and higher powers.)

Pythagoras tells us is that if we add the squares of the two shorter sides, we get the square of the longest side. Putting this in algebraic terms:

$$a^2 + b^2 = c^2$$

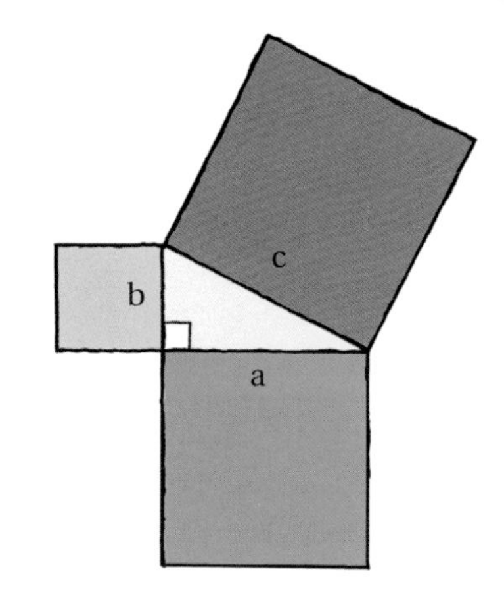

There is a famous geometric picture, which illustrates this fact (see right).

The point is that the areas of the two smaller squares (a^2 and b^2) together add up to the area of the largest square (c^2). But what does this mean in practical terms – how can we use it?

Calculating with Pythagoras

How can we calculate actual lengths using Pythagoras' theorem? Let's take a simple example: a right-angled triangle which is 2cm tall and 2cm wide (so it's a 2×2 square, cut in half diagonally).

What is the length of the third side (the diagonal of the square)? Pythagoras' theorem tells us that $a^2 + b^2 = c^2$, and in this particular example $a = b = 2$, and c is what we want to work out. Substituting $a = b = 2$ in the formula, we get $2^2 + 2^2 = c^2$. Working out this out, we find that $c^2 = 4 + 4 = 8$.

How to we calculate c if we know c^2? The answer is given by the *square root* that we met in *Root and logs*. It must be that $c = \sqrt{8}$. Typing this into a calculator we get the answer: $c = 2.83$cm, to two decimal places. (Notice that, even though the numbers in the question were whole numbers, the answer isn't. This is very common.)

NOW HAVE A GO AT QUIZ 1.

Calculating the shorter sides

Sometimes we might already know the longest side of the triangle, and want to know one of the shorter ones. For instance, suppose a right-angled triangle is 24cm wide with a hypotenuse of 25cm. The question we want to answer is: how high is it? If we call the height a, then the theorem tells us that:

$$a^2 + 24^2 = 25^2$$

Working these squares out, this becomes:

$$a^2 + 576 = 625$$

Using a technique we met in *Equations*, we subtract 576 from both sides, to get:

$$a^2 = 625 - 576 = 49$$

Finally, we take the square root of both sides: $a = \sqrt{49}$, which we might recognize (without using a calculator) as giving $a = 7$cm.

TRY USING PYTHAGORAS' THEOREM THIS WAY IN QUIZ 2.

Distance and Pythagoras

What use is Pythagoras' theorem? It is not as if we see right-angled triangles every day . . . or do we?

Actually, it is quite common to want to know the length of a diagonal line when we already have information about horizontal and vertical distances. In such situations, we can get help from hidden right-angled triangles.

For example, suppose a rectangle is 5cm wide and 12cm long. How long is its diagonal? The answer is not obvious, but Pythagoras comes to the rescue, because when we draw in the diagonal, we split the rectangle into two right-angled triangles that have the same length and width as the rectangle. The diagonal of the rectangle is the hypotenuse of the right-angled triangle. So we can use Pythagoras' theorem. Whatever the length of the diagonal is, call it c, it must be true that:

$$5^2 + 12^2 = c^2$$

Working this out, we get $c^2 = 25 + 144 = 169$. So we want a number which, when squared, gives 169. With or without a calculator, we can identify the answer as $c = 13$cm.

Here is another example. Suppose a church tower is 100 metres tall. My wife waves to me from the top, while I sit sipping coffee in a café on the other side of the piazza below, 75 metres away from the base of the tower. How far is my wife from me?

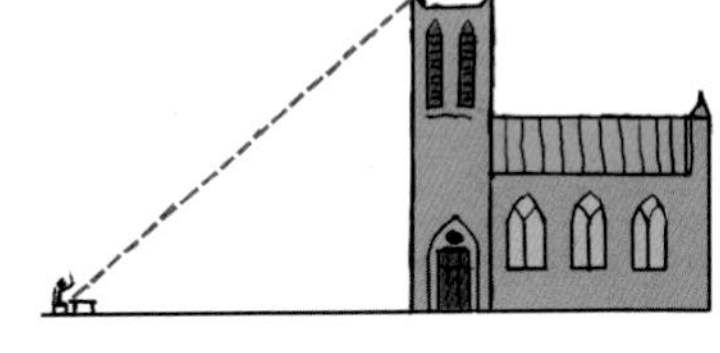

There is a right-angled triangle hidden in this scenario. One edge is provided by the line from me to the base of the church tower, and another is the church tower itself. These are 75m and 100m respectively, and crucially these two are at right angles to each other (assuming that the tower goes straight up and down, not like the leaning tower of Pisa). The third side (the hypotenuse) is now the length we want, because this has me at one end and my wife at the other. Again we call this length c, and Pythagoras' theorem assures us that:

$$75^2 + 100^2 = c^2$$

To solve this, we first work out $75^2 = 5625$ and $100^2 = 10{,}000$. Adding these together, we find that $c^2 = 15{,}625$. So the final step is to take the square root of 15,625, and find that $c = 125$m.

TRY USING PYTHAGORAS' THEOREM IN QUIZ 2.

A beautiful proof!

I hope you have seen that Pythagoras' theorem can be genuinely useful for calculating lengths, whenever a right-angled triangle can be found. But why should it be true? This theorem has been proved in many different ways, probably more than any other theorem in the history of mathematics. Each generation of mathematicians provides a new proof. In 1907, Elisha Loomis assembled a collection of 367 different proofs, but the total number is much higher.

One particularly neat proof was discovered by Bhaskara in 12th-century India. Start with a right-angled triangle with sides of length a, b and c, where c is the hypotenuse as usual. Now we form two large squares, each with side length $a + b$.

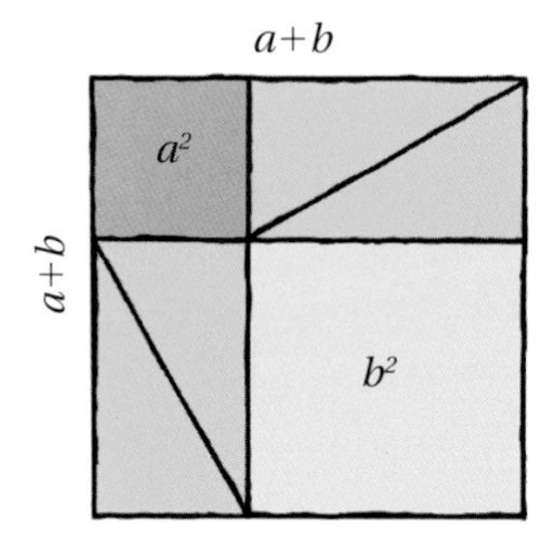

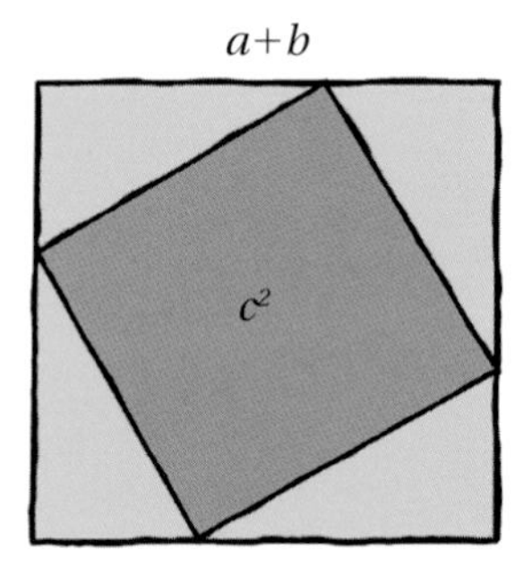

The first large square is divided up inside into four copies of the original triangle, together with one smaller square of area a^2 and another square of area b^2. The second large square is divided up differently, into four copies of the original triangle and one smaller square of area c^2. The crucial observation is that the areas of the two large squares are the same (obviously, because they are identical). In each case, there are four copies of the right-angled (a, b, c) triangle inside; although these are arranged differently, the total area of these four triangles must be the same. That means that the area that is left over in each of the two large squares must be the same. In the first square the area left over is $a^2 + b^2$ and in the second square the area left over is c^2 so these two quantities must be equal, that is, $a^2 + b^2 = c^2$.

PLAY AROUND WITH THIS PROOF IN QUIZ 4.

Such a wonderful theorem is worth spending some time thinking about – even if it's not immediately obvious!

Pythagorean triples

Perhaps the simplest right-angled triangle is half a square. If the square is 1cm wide, then Pythagoras' theorem tells us that the hypotenuse c satisfies $c^2 = 1^2 + 1^2$ so $c^2 = 2$, and it must be that $c = \sqrt{2}$cm. It is slightly inconvenient that this is not a whole number. In fact, it is even worse. Like the number π (see *Circles*), $\sqrt{2}$ is an *irrational* number, meaning that it cannot be written exactly as a fraction, or even as a recurring decimal.

This is what usually happens. For example, if you draw a right-angled triangle with shorter sides 1cm and 2cm, the hypotenuse will be $\sqrt{5}$cm, which again is irrational.

Occasionally, however, we do get a whole number as the answer. If the two shorter sides are 3cm and 4cm, then the hypotenuse is exactly 5cm.

These situations, where all three sides of a right-angled triangle are whole numbers, are known as *Pythagorean triples*. The first one is 3, 4, 5. If we double all these lengths, we get another: 6, 8, 10. Multiplying all the lengths by other numbers will also produce further Pythagorean triples.

EXPLORE PYTHAGOREAN TRIPLES IN QUIZ 5.

Apart from multiples of 3, 4, 5, the next Pythagorean triple is 5, 12, 13. Pythagorean triples are rather mysterious and unpredictable things!

Sum up *Pythagoras' theorem is one of the great geometrical theorems. It is just a question of putting it to good use!*

Quizzes

1 Calculate the hypotenuse (the longest side) of right-angled triangles which have these shorter sides.

a 3 miles and 4 miles

b 1cm and 1cm

c 2 metres and 3 metres

d 5mm and 6mm

e 8km and 15km

2 Calculate the missing side! In each case c is the hypotenuse, and a and b are the shorter sides.

a a = 1mm and c = 3mm

b b = 2 miles and c = 4 miles

c a = 1cm and b = 9cm

d b = 7km and c = 11km

e a = 11 inches and c = 61 inches

3 Calculate these distances (to one decimal place). In each case, you will need to identify a right-angled triangle, and spot which side is the hypotenuse.

a A rectangular room is 4 metres long and 3 metres wide. What is the distance from one corner to the diagonally opposite corner?

b A flag is 38cm wide and 56cm long. A black stripe runs diagonally from one corner to another. How long is the black stripe?

c An aeroplane flies directly over my friend's house at an altitude of 10km. I am 1km away from her house. How far is the aeroplane from me?

d A football pitch is 119m long and 87m wide. If I run from one corner to the diagonally opposite corner, how far have I run?

e A submarine is 300m east of a ship on the sea surface, and is at a depth of 400m. How far apart are the ship and the submarine?

4 **a** Draw a right-angled triangle. Now try to recreate Bhaskara's proof using your triangle!

b A right angle triangle has sides of length $a = 3$cm, $b = 4$cm, and $c = 5$cm. Find its area (see chapter on Triangles) and find the areas of the four squares used in Bhaskara's proof. Check that the proof works!

c In a Bhaskara proof, the four squares have areas of 6.25cm^2, 36cm^2, 42.25cm^2, and 72.25cm^2. What are the dimensions of the right-angled triangle?

d In a Bhaskara proof, the largest outer square has an area of 529cm^2 and the smallest inner square has an area of 64cm^2. What are the dimensions of the right-angled triangle?

5 Use Pythagoras' theorem to complete these Pythagorean triples.

a 7, 24, ? **b** 8, 15, ? **c** 9, 40, ?
d 11, ?, 61 **e** ?, 35, 37

Trigonometry

- *Understanding what trigonometry is all about*
- *Knowing how to use trigonometry to understand triangles*
- *Deciding which of sin, cos and tan to use*

So far, we have seen that triangles of any shape have one thing in common: their three angles add up to 180° (see* Triangles*). We have also met the famous Pythagoras' theorem, which relates the three lengths of any right-angled triangle. What we don't yet have is anything which relates the two: something which allows us to calculate the lengths in a triangle if we know the angles, and vice versa. This is the subject of* trigonometry*, known as* trig *to its friends, a word which literally means 'triangle-measuring'.

Similar triangles

Let's have a look at some right-angled triangles. Suppose we fix one of the angles to be 60°, and we want to know what the lengths of the sides are. Notice that we now know two of the angles: the right angle (90°) and the angle of 60°. Since the three angles must add up to 180°, it must be that the third and final angle is 180° − 90° − 60° = 30°.

Now, none of this determines the lengths of any of the sides. It could be that the triangle is 100 miles long or just 0.5mm. Yet, when we look at different triangles with this same arrangement of angles, there is something very striking about them.

Although these triangles are different *sizes*, when we look at them, they are all clearly the same *shape*. (The technical term for being the same shape,

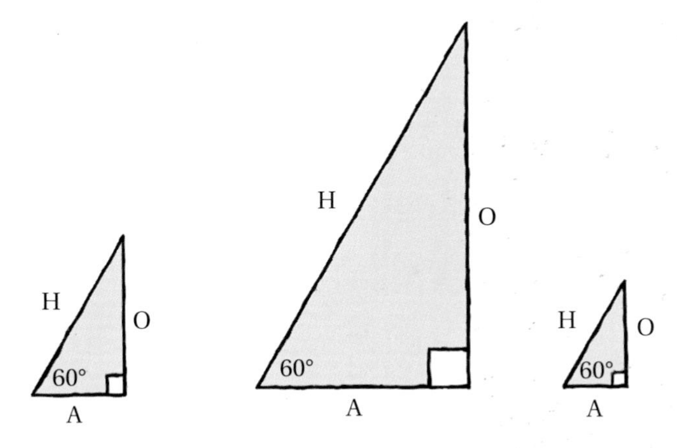

but not necessarily the same size, is being *similar*.) The moral is: if we know all three angles then, as soon as we fix the length of *one* of the sides, in order for the triangle to have the right shape (and thus have the correct angles), the other two sides are automatically determined. There is no further room for manoeuvre. This is the key idea behind trigonometry. In the rest of the chapter we will put some technical meat on these bones.

Keeping things in proportion

What does it really mean for two triangles to be the same *shape*, but not necessarily the same *size*? It means that their three angles must match. That's fine, but what does it mean for the lengths of the sides? Well, however long or short they are, the three sides in the two triangles must be in the same *proportion* to each other.

In the 60° triangles above, it turns out that the shortest side (the one at the bottom) is always half the length of the longest side (in the terminology of *Pythagoras' theorem* that's the *hypotenuse*). So, if we are presented with a right-angled triangle which also contains an angle of 60°, and we know that the hypotenuse is 4cm, then it automatically follows that the shortest side must be 2cm long.

It is customary to name the three sides of the triangle H (for *hypotenuse*), A (for the side *adjacent* to the angle we are focusing on, in this case the 60° angle) and O (for the side *opposite* that angle). What the last paragraph says is that, in any such triangle, it will always be true that $\frac{A}{H} = \frac{1}{2}$.

Now, the number $\frac{1}{2}$ is specific to the angle of 60°, but the line of reasoning is not. For any angle that can occur in a right-angled triangle, there is a fixed number to which $\frac{A}{H}$ will always be equal. Usually these numbers are less neat than $\frac{1}{2}$. In the case of 45°, for example, it will always be true that $\frac{A}{H} = \frac{1}{\sqrt{2}}$. In the case of 30°, it will always be true that $\frac{A}{H} = \frac{\sqrt{3}}{2}$.

The word we use for this number is *cosine*. We say that the cosine of 60° is $\frac{1}{2}$, the cosine of 45° is $\frac{1}{\sqrt{2}}$, and the cosine of 30° is $\frac{\sqrt{3}}{2}$. Cosine is usually abbreviated to *cos*, and we would write: $\cos(60°) = \frac{1}{2}$, and $\cos(45°) = \frac{1}{\sqrt{2}}$, and $\cos(30°) = \frac{\sqrt{3}}{2}$.

What is the relationship between the angle and the corresponding number, its cosine? How can we calculate the cosine? Here, my advice is simple:

leave it to your calculator! This is not like addition or multiplication, or even like logarithms: there is no simple way to compute the cosine of an angle by hand. But your calculator does have a button which can do it for you: [cos]. So, to calculate the cosine of 18°, you would type [cos] [1] [8] [=].

Now have a go at Quiz 1

Warning! Make sure the mode is set to *degrees* not *radians*, otherwise these calculations will come out wrong! (You may need to consult your calculator's handbook to do this.)

Calculating lengths from angles

Let's suppose that we have a right-angled triangle. We know that one of its angles is 18°, and we know that the hypotenuse (H) is 4cm long. This determines the length of the adjacent side. But what is that length? Let's calculate it. Whatever the value of cos(18°), it must be true that $\frac{A}{H} = \cos(18°)$. We also know that $H = 4$cm. So $\frac{A}{4} = \cos(18°)$.

We want to know A, but this equation gives us $\frac{A}{4}$. So we need to multiply by 4. The rules of equations apply here, specifically the command to keep the equation balanced by doing the same thing to both sides (see *Equations*). Multiplying both sides by 4, therefore, we get $A = 4 \times \cos(18°)$.

This we can type straight into a calculator: [4] [×] [cos] [1] [8] [=], giving an answer of 3.8, to one decimal place. So the length of the adjacent side is 3.8cm. Notice that it was worth resisting calculating cos(18°) itself, until the very last moment!

The argument is slightly different if we are faced with a right-angled triangle where we know an angle and the length of the adjacent side, and we want to work out the hypotenuse. Suppose a right-angled triangle has an angle of 75°, with an adjacent side of 6 metres. This means that $\frac{A}{H} = \cos(75°)$, but in this case we know that $A = 6$m. So $\frac{6}{H} = \cos(75°)$. We want to get H on its own on one side of the equation, so we start by multiplying both sides by H. This gives us $6 = H \times \cos(75°)$. Now, cos(75°) is just a number, even if we don't know its value yet, so we can divide both sides by that to get $H = \frac{6}{\cos(75°)}$.

This is now something we can tap into our calculator: [6] [÷] [cos] [7] [5] [=], which gives an answer of 23.2 metres to one decimal place.

Practise calculations like this in Quiz 2.

SOH-CAH-TOA!

So far, we have focused on the relationship between the adjacent side (A) and the hypotenuse (H), as expressed by the *cosine* of the angle between them, which we might call x. But there are similar relationships between the other pairs of sides too. Between the opposite side (O) and the hypotenuse (H), we have the *sine*, or *sin* for short: $\sin x = \frac{O}{H}$. With this, if we know an angle and the length of the opposite side, we can calculate the hypotenuse H (and vice versa).

Similarly, between the opposite side (O) and the adjacent side (A) we have the *tangent*, or *tan* for short, defined by $\tan x = \frac{O}{A}$.

Armed with the three *trigonometric functions*, sin, cos and tan, it is possible to work out a lot of information about triangles. But first we need to remember which is which! There are various mnemonics to help us remember these; the shortest is 'SOH-CAH-TOA':

$$\text{Sin } x = \frac{O}{H}$$
$$\text{Cos } x = \frac{A}{H}$$
$$\text{Tan } x = \frac{O}{A}$$

(Soh-Cah-Toa is phoney Chinese, very roughly translating as 'Big Foot Aunt' in Mandarin.)

The power of trig!

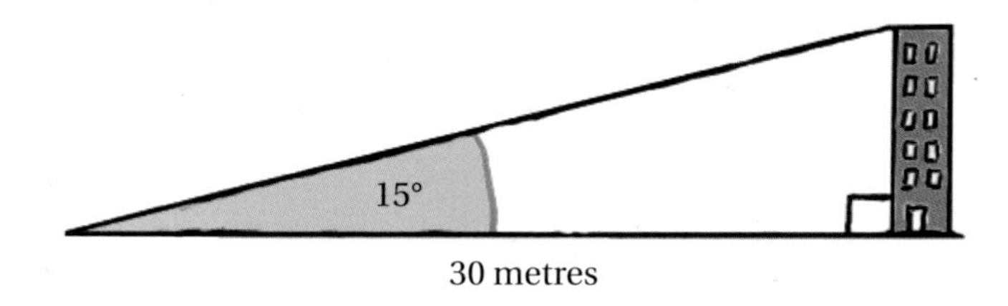

Let's see the power of trigonometry. Suppose a tower is 30 metres away from me. When I look at the top of the tower, the angle of elevation from me to the top is 15°. The question is: how tall is the tower?

There's a right-angled triangle here, in which the three corners are: the point where I am standing, the base of the tower and the top of the tower. So, we can use trigonometry. What is more, we know one of the angles: 15°. The first thing to decide, then, is which trigonometric function (cos, sin or tan) we want to use. We know the adjacent side (A): 30m. What we want to know is the opposite side (O): the height of the tower. So, SOH-CAH-TOA tells us that the right choice is the *tan* function because that's the one that involves O and A.

TRY APPLYING THIS YOURSELF IN QUIZ 3.

The TOA formula tells us that $\tan(15°) = \frac{O}{30}$, so $O = 30 \times \tan(15°)$. Typing this into a calculator [3] [0] [×] [tan] [1] [5] [=], we get an answer of 8.04 metres to two decimal places.

All the examples we have seen so far fit a similar pattern:

- We have a right-angled triangle.
- We know one of its angles (apart from the right angle), as well as the length of one of its sides.
- We want to calculate the length of one of its other sides.
- Pick the appropriate trigonometric funtions (sin, cos and tan), using SOH-CAH-TOA.
- Plug the numbers we have into the formula.
- Rearrange the equation.
- Calculate the answer.

Doing trigonometry backwards!

There is another way in which trigonometry can be useful, which doesn't follow the above pattern. If we know the lengths of *two* of the sides of a right-angled triangle, then we can calculate the remaining angles (the ones that are not the right angle). Of course, we can also calculate the third side, using that old standby, Pythagoras' theorem.

For instance, suppose we know that a triangle has a hypotenuse of 5m, and one of its other sides is 4m long. Let's say we want to know the size of the angle opposite the 4m side; call the angle x. SOH-CAH-TOA tells us that $\sin x = \frac{4}{5}$.

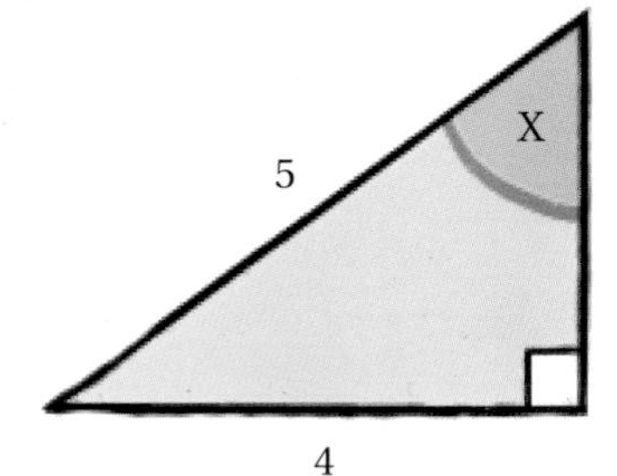

How we can use this to find out x? Well, what we need to do to the left of the equation is 'undo' sine, somehow, and on the basis of always doing the same thing to both sides of the equation, we need to 'undo sine' on the right as well.

This may be starting to sound like nonsense, but actually it is perfectly meaningful. *Sine* is a rule which takes in a number representing an angle, and spits out another number (representing the ratio of two sides, $\frac{O}{H}$ in the related right-angled triangle). There is another rule which does the same thing backwards: it takes in the value of $\frac{O}{H}$ and spits out the corresponding angle. This function is called *inverse sine*, and is written $\sin^{-1}$ (or sometimes 'arcsin').

Inverse sine can be found on most calculators by pressing SHIFT followed by sin. So, to complete the example above, we have arrived at $\sin x = \frac{4}{5} = 0.8$ (converting the fraction to a decimal: see *Fractions* if you have forgotten how to do this). Now we need to take the inverse sine of both sides: $x = \sin^{-1}(0.8)$. All we need to do now is calculate this using a calculator: $\sin^{-1}$ 0 · 8 =, giving a final answer of 53°, to the nearest degree.

Have a go at using them in Quiz 4

There are also inverse cosine and inverse tangent rules, which relate to cosine and tangent in exactly the same way.

Sum up *Trigonometry is the science of measuring triangles, and its three main planks are sin, cos and tan. So long as you have a calculator, and can remember SOH-CAH-TOA, trigonometry is not too hard at all!*

Quizzes

1 Use a calculator to find these.

a $\cos(72°)$ **b** $\cos(61°)$ **c** $\cos(11°)$ **d** $\cos(59°)$ **e** $\cos(28°)$

2 Calculate the length of the adjacent side, when the hypotenuse (H) and the given angle (x) are as follows.

a $H = 6$ miles, $x = 54°$ **b** $H = 3$cm, $x = 12°$
c $H = 3$ metres, $x = 77°$ **d** $H = 12$ nanometres, $x = 45°$
e $H = 45$mm, $x = 34°$

3 In each case below, x is an angle in a right-angled triangle, A is the adjacent side, O is the opposite side and H is the hypotenuse. Choose the correct trig function, and calculate the unknown length to one decimal place.

a $x = 35°$, $O = 2$ miles, what is H?
b $x = 22°$, $A = 5$cm, what is O?
c $x = 70°$, $H = 1$km, what is O?
d $x = 41°$, $A = 67$mm, what is H?
e $x = 12°$, $O = 1$ micrometre, what is A?

4 In each case below, the lengths of two sides of a right-angled triangle are given. Choose the correct inverse trig function to discover the angle that is opposite the side O and adjacent to the side A.

a $A = 3$cm, $H = 5$cm **b** $H = 6$mm, $O = 1$mm
c $A = 5$ miles, $O = 7$ miles **d** $O = 12$km, $H = 14$km
e $H = 55$ metres, $O = 24$ metres

- *Understanding what axes are*
- *Interpreting coordinates*
- *Knowing how to plot points on a graph*

***So far, we have met a variety of geometrical objects and techniques. To take the subject further, we need a feeling for the mathematical* space *where geometry takes place, the background on which our lines and shapes are drawn and studied. In two dimensions, this role is played by the* plane, *essentially an idealized piece of paper. Unlike any paper in the physical world, however, it is perfectly smooth, flat, and – well – plain. And it is* infinite, *extending forever in all directions. Of course, no physical piece of paper can hope to achieve this! How do you know where you are on an infinite piece of paper? The answer comes in the form of* Cartesian coordinates, *named after the French philosopher René Descartes, also known by his pen-name,* Cartesius.**

Coordinates are familiar to anyone who has ever read a map, and Cartesian coordinates work in a very similar way, as a map of the infinite plane.

Introducing the axes

The starting point for Cartesian coordinates is as follows: there are two lines drawn across the plane. One runs horizontally from left to right, and the other runs vertically up and down. These are known as the horizontal and vertical *axis* respectively (or as the *x-axis* and *y-axis*, for reasons we shall come to in the next chapter).

Each of the two axes has numbers along it. The horizontal axis is identical to the *number line* that we met in *Negative numbers and the number line*. In the middle is the number 0. To its left are the negative numbers: -1, -2, -3, . . . To its right are the positive numbers: 1, 2, 3, . . . (We are not only interested in the whole numbers here: between 1 and 2 is a full range of fractions and decimals.)

The vertical axis, as its name suggests, runs from down to up instead of from left to right, but is otherwise identical to the horizontal axis. These two lines meet in only one place: where they are both 0. This special point is known as the *origin*.

This pair of axes forms a cross in the middle of the plane. But they can be used to locate *every* point on the plane. This is the idea of *Cartesian coordinates*.

Descartes' infinite map

A pair of Cartesian coordinates consists of two numbers, usually written inside brackets, and separated by a comma like this: (4, 5). This pair of numbers pinpoints a position on the plane. How? These are the rules:

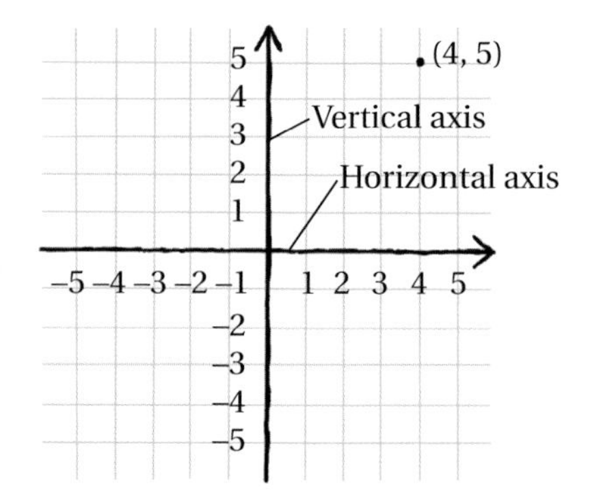

- The first coordinate says how far *left or right* the point is from the origin. In this case it is 4, so the point will be in line with the 4 on the horizontal axis.
- The second coordinate (5 in this case) tells us how far *up or down* the point is from the origin. So it must be in line with the 5 on the vertical axis. The most common type of confusion with coordinates is getting is getting (4, 5) muddled up with (5, 4). You need to remember that the first number in the pair counts from left to right, and the second from down to up. This chapter's golden rule is a mnemonic for remembering the correct order of the coordinates. If you are used to reading ordinary map references, you will find this easy as the system is essentially the same.

IDENTIFY AND PLOT POINTS IN QUIZZES 1 AND 2.

We also say that the point (4, 5) has an *x-coordinate* equal to 4, and a *y-coordinate* equal to 5. (Now we can remember that the coordinates appear in alphabetical order: we just have to remember which axis is which!) The origin has coordinates (0, 0).

Distance

When we have two points on the plane, an obvious question to ask is how far apart they are. How can we get this information from coordinates?

The answer comes from that bedrock of geometry, Pythagoras' theorem, which we discussed earlier. Let's look at the two points $(-1, -2)$ and $(3, 1)$. The first step is to work out the horizontal and vertical distances between the points.

The horizontal coordinate is -1 at the first point and 3 at the second, which is an increase of 4. That is the horizontal distance between the points. Similarly, the first vertical coordinate is -2 and the second is 1, which gives a vertical difference of 3. That is the vertical distance between the points.

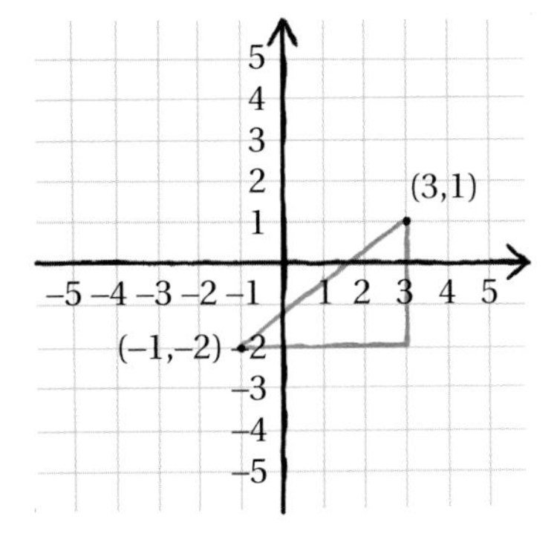

These are the horizontal and vertical sides of a right-angled triangle, so Pythagoras' theorem tells us how to use these two figures to work out the distance between the two points. The length we want is that of the hypotenuse. If we call that c, then $c^2 = 3^2 + 4^2$, which gives an answer of $c = \sqrt{25} = 5$.

Breaking this down, the method for calculating the distance between two points is as follows:

- Calculate the horizontal and vertical distances between the two points.
- Use Pythagoras' theorem to calculate the direct distance.

CALCULATE DISTANCES IN QUIZ 3.

Sometimes there is a short cut. Consider the two points (1, 2) and (5, 2). In this case the horizontal distance is 4, and the vertical distance is 0. So, in this case, the actual distance is the same as the horizontal distance: 4.

Into three dimensions!

The plane is the right setting for studying circles, triangles, and so on. But what about the likes of spheres and cubes? There is no room in the plane for objects like this, which have depth as well as length and height.

This is the moment we need to step from two dimensions into three. The system of 3-dimensional coordinates works very similarly to the 2-dimensional one. The difference is that there is an extra axis. Before we had only an x-axis and a y-axis. These are now joined by a z-axis. Usually the three axes are arranged as shown in the diagram.

This time, any point in 3-dimensional space can be represented by a triplet of numbers such as (1, 2, 3), which give the point's position along the x-, y-

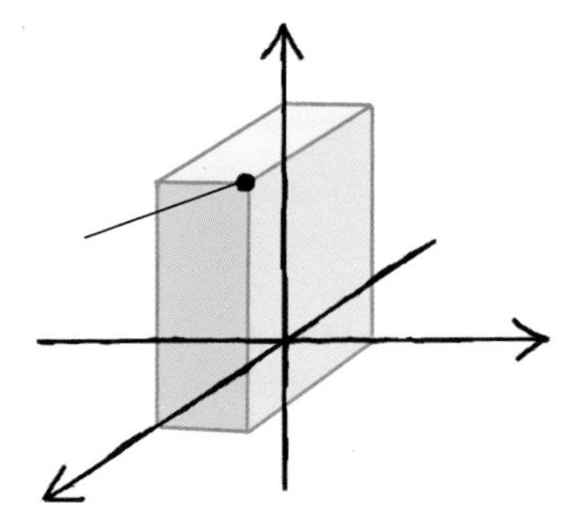

and z-axes respectively. The biggest challenge here is artistic! It helps to draw in guidelines parallel with the axes, to show where your point really is.

Pythagoras in three dimensions

Finding the distance between two points in 3-dimensional space is almost the same as in two. Suppose we want to find the distance between the points (1, 2, 3) and (3, 5, 7). You might be able to guess the method! First find the distance between the two points along each coordinate axis. The x-distance is $3 - 1 = 2$, the y-distance is $5 - 2 = 3$, and the z-distance is $7 - 3 = 4$. Then, following the method that we used in two dimensions, we square each of these distances, add them up and take the square root: $\sqrt{2^2 + 3^2 + 4^2} = \sqrt{29} = 5.39$, to two decimal places.

PLOT SOME POINTS IN 3-DIMENSIONAL SPACE IN QUIZ 4.

Although this method is a straightforward adaptation of the 2-dimensional version, you might wonder exactly what is going on. After all, we seem to be using Pythagoras' theorem, but with three numbers being squared and added up to find a fourth. What can this mysterious four-sided triangle be?

It is not so mysterious really! If we ignore the z-dimension for a moment, then the distance between the two points is $\sqrt{2^2 + 3^2}$, using Pythagoras' theorem just as before. We might call this the $x{-}y$-distance.

But the $x{-}y$-distance (that's $\sqrt{2^2 + 3^2}$) and the z-distance (4) form the first two sides of another right-angled triangle, and what we want to know is the hypotenuse. So again we square them, add them together, and then take the square root. But squaring $\sqrt{2^2 + 3^2}$ gives us back $2^2 + 3^2$, so the answer will be $\sqrt{2^2 + 3^2 + 4^2}$, as expected!

CALCULATE SOME DISTANCES IN THREE DIMENSIONS IN QUIZ 5.

Sum up *Axes and coordinates are to geometers what a map and compass are to mountaineers!*

Quizzes

1 *What are the coordinates of each of the points a to e?*

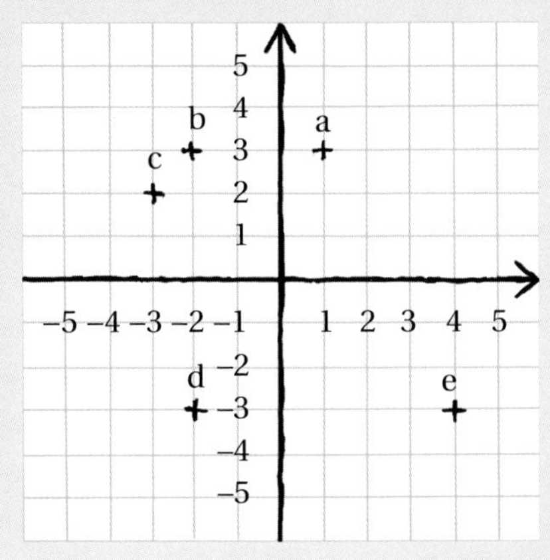

2 *Draw and label a pair of axes, and plot these points.*

a $(1, 2)$
b $(-2, 1)$
c $(3, -3)$
d $(-3, -2)$
e $(-2\frac{1}{2}, -2\frac{1}{2})$

3 *Calculate the distance between each of these pairs of points.*

a $(1, 2)$ and $(5, 5)$
b $(0, -4)$ and $(5, 8)$
c $(-1, 1)$ and $(1, -1)$
d $(-2, -2)$ and $(-2, 1)$
e $(-2, -3)$ and $(6, 7)$

4 *Draw a set of x-, y- and z-axes, and plot these points.*

a $(1, 1, 1)$
b $(3, 2, 1)$
c $(-1, 3, 2)$
d $(2, -2, 1)$
e $(0, 3, -3)$

5 *Find the distance between the point $(2, 2, 2)$ and each of the points in quiz 4 above.*

Graphs

- *Knowing how to plot a graph*
- *Finding the equation of a line*
- *Understanding how geometry and algebra are related*

In the last chapter, we saw how to equip the plane with axes and coordinates. This now gives us two ways of looking at it: the first comes directly from the lines and shapes that we can draw, and the second is numerical, coming from the coordinates of the points within those lines and shapes. How these two perspectives are related is a very deep question, one that mathematicians continue to study today. In this chapter we look at the ways to move between the two, in the particular case of straight lines.

Let's start by looking at the collection of all the points which have coordinates like this: (0, 0), (1, 1), (88, 88), (2, 2), (−1.5, −1.5), etc. The numerical pattern is, I hope, clear. These are the points whose two coordinates are equal.

But what pattern do they make when we draw them on the plane? There is only one way to find out!

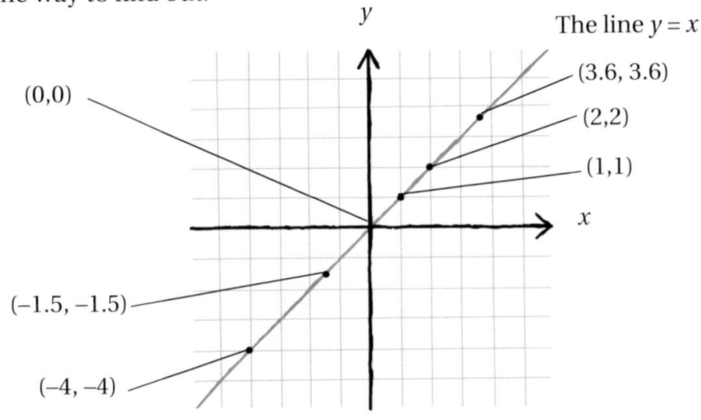

What we have here is a straight line. What more can we say about it? Well, to start with, we could ask about its relationship with those two other lines: the axes. Where does it cross them? The answer is that it crosses each of them at the same place: the origin (0, 0). What else might we want to say? On closer inspection, we can see that our new line is at an angle of 45° to the horizontal.

From a geometric perspective, these pieces of data, that it is a straight line, which passes through the origin, and has an angle of 45°, pin it down uniquely. If you want to draw something fitting this description, there is only one possible line you can draw.

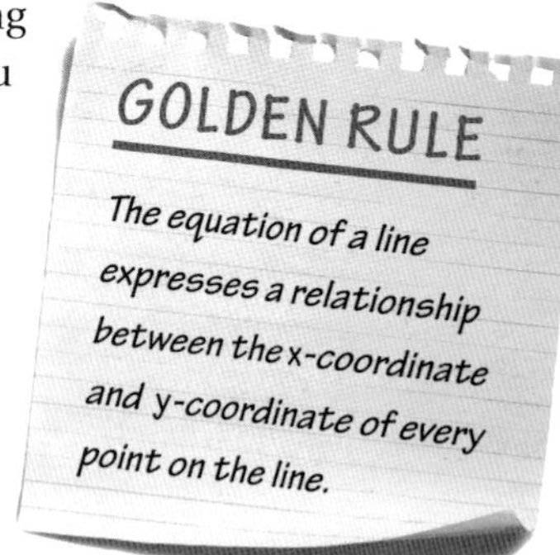

The equation of a line

When we have a pair of coordinates, it is traditional to refer to the first one as x, and the second as y, like this: (x, y). So the point (8, 9) has an x-coordinate of 8 and a y-coordinate of 9. (This is the reason why the horizontal axis is also known as the *x-axis*, and the vertical as the *y-axis*.)

What is the purpose of this extra jargon? Let's return to the line that comprises all the points like (1, 1), (88, 88) and (−1.5, −1.5). What is the defining characteristic of this line, in numerical terms? It is that the two coordinates x and y are always equal. We can write this as $y = x$.

This formula $y = x$ is known as the *equation* of this particular line. The relationship between lines and their equations is the main topic of this chapter.

From lines to equations

Let's have another example. Look at all the points of this form: (0, 1), (1, 2), (−4, −3), (88, 89), and so on. What is the pattern here? In each case the second coordinate (the y-coordinate) is 1 more than the first (the x-coordinate). We can write this as $y = x + 1$. This is the equation of this line.

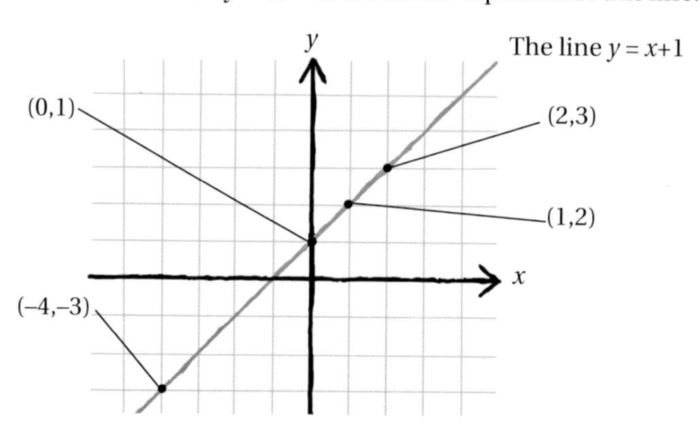

We could equally well think of this as saying that the first coordinate is 1 less than the second, and write the equation as $x = y - 1$. This is perfectly correct, but it is traditional to express the y-coordinate in terms of the x-coordinate. So the equation of a line usually begins '$y = \ldots$'

TRY THIS YOURSELF IN QUIZ 1.

The only exceptions are lines like the one comprising all the points (4, 0), (4, 1), (4, 2), and so on. When we draw it on the graph, this comes out has a vertical line, and it has the equation $x = 4$. (There is no constraint on what value y can take: (4, 1,000,000) is another point on this graph.)

Here is another example. Take the line comprising points like these: (0, 0), $(1, \frac{1}{2})$, (2, 1), $(3, 1\frac{1}{2})$, and so on. The rule here is that the y-coordinate is always half the value of the x-coordinate. So the equation is $y = \frac{1}{2}x$.

Plotting graphs: from equations to lines

Let's turn the whole thing on its head. Suppose we are given an equation, such as $y = 2x - 1$, and we are set the challenge of drawing the corresponding line. How can we start?

We need to find some points that lie on the line. In this context, knowing a point means knowing its coordinates. The type of question we need to ask is as follows: when $x = 1$, what is y? There is nothing magic about the number 1 here, any other number would have done: when $x = 0$,what is y? When $x = 4$, what is y? And so on.

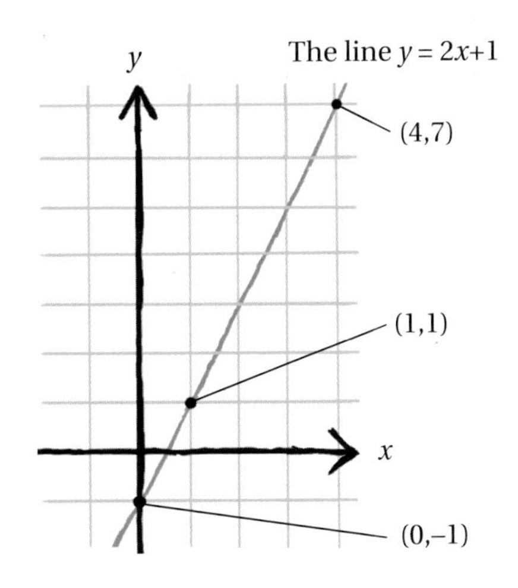

The way to answer such questions is to *substitute* the chosen value of x into the equation, and see what emerges for y. When $x = 1$, then according to the equation, $y = 2 \times 1 - 1$, which is 1. So we have the coordinates of our first point on the line: (1, 1). We can draw a dot or cross on our graph to mark this point.

When $x = 0$, the equation tells us that $y = 2 \times 0 - 1$,which comes out as $y = -1$. So another point on the line is $(0, -1)$, which we can add to the graph.

At this moment we could stop, as two points are enough to determine the whole line: there is only one straight line which we can draw, which goes through both of these points. But it is good practice to calculate at least one more point on the line, just to check that the three really do lie in a straight line, and we have not made a mistake. (It's always a good plan to guard against human error.) When $x = 4$, the equation tells us that $y = 2 \times 4 - 1 = 7$. So a third point is $(4, 7)$, which we also add to the graph. We can finish off by connecting these three points with a line.

HAVE A GO AT THIS YOURSELF IN QUIZ 2.

Steepness and gradients

So far, we have seen two perspectives on straight lines: a visual, geometric perspective and an algebraic one, involving numbers and equations.

If we compare the lines we have seen so far in this chapter, we might say that one is *steeper* than another. This notion of *steepness* is critical to the geometric viewpoint. As we shall see, it also ties in very neatly to the equation-based approach.

First we need to make the idea of steepness precise. Earlier in the chapter, I described a line as being *at 45° to the horizontal*. In doing so, I used an angle to measure a line's steepness. This works perfectly well but, for the most part, we prefer a different measure, called the *gradient*.

When driving along a mountain road, you might have seen a sign which says 'Hill ahead 1:4' or 'Hill ahead, 25%'. These expressions "1:4" and '25%' are measures of the gradient of the hill. What they both mean, in precise

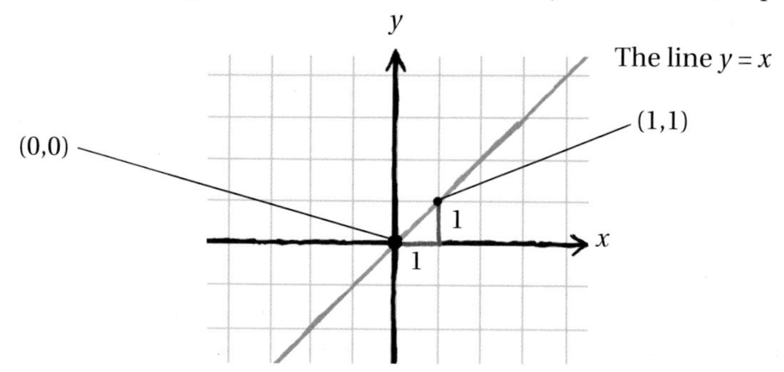

terms, is that for every 4 metres you travel horizontally, you will travel 1 metre vertically.

This is what the gradient is: the amount by which a line (or hill) rises vertically per unit of horizontal distance. In mathematics, this is usually expressed as a fraction rather than a ratio or a percentage. So a geometer would write the gradient of that hill as $\frac{1}{4}$. The general notion of gradient is $\frac{\text{Vertical difference}}{\text{Horizontal difference}}$.

Let's calculate this for some examples, starting with the line $y = x$, which we saw above. Pick any two points on the line, say (0, 0) and (1, 1). Between these points, there is 1 unit of horizontal distance: this can be read from the difference between the first coordinates of the two points: from 0 to 1. What is the vertical difference between them? The answer is the same, 1 again, as the difference between the y-coordinates is also 1. So the gradient is $\frac{1}{1} = 1$.

Let's take another example: $y = 2x - 1$. Pick two points on it, say $(0, -1)$ and (1, 1). Again the horizontal difference between them is 1 (this is the difference in their first coordinates). The vertical difference between them is 2, which we can see either from the picture, or by reading from their second coordinates, since $1 - (-1) = 1 + 1 = 2$. Remember the rules of subtracting negative numbers from an earlier chapter! So the gradient of this line is $\frac{2}{1} = 2$. It makes sense that this should have a larger gradient than the previous example, since the line does indeed look steeper.

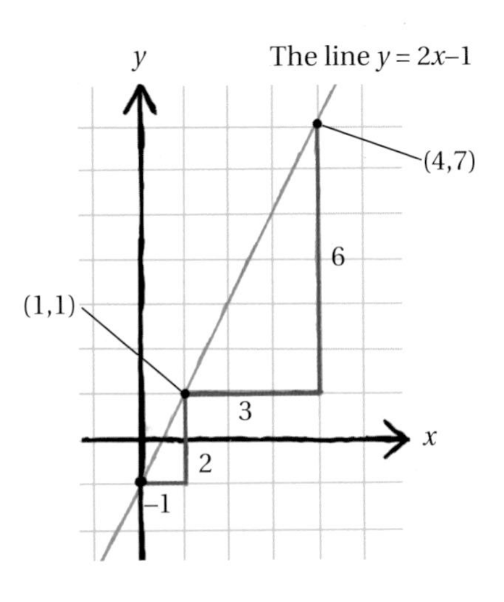

You might worry that the value we get for the gradient seems to depend on the choice of the two points on the line. What if, in the last example, we had instead picked (1, 1) and (4, 7)? This time the horizontal difference is 3 (since $4 - 1 = 3$) and the vertical difference is 6 (since $7 - 1 = 6$). So the gradient comes out as $\frac{6}{3} = 2$. It is no coincidence that this is the same as the previous answer: it always will be. Unless we mess up the calculation, the gradient does not depend on the choice of points used to calculate it. A straight line has the same steepness, wherever you calculate it.

A third example: the line $y = \frac{1}{2}x$. We saw above that two points on this line are $(1, \frac{1}{2})$ and $(2, 1)$. The horizontal difference between these is 1 (since $2 - 1 = 1$). The vertical difference is $\frac{1}{2}$ (since $1 - \frac{1}{2} = \frac{1}{2}$). So the gradient is $\frac{1}{2}$.

Downhill all the way

In all the lines we have seen so far, as you move rightwards along the line, you also move upwards. Those lines go *uphill*, if you like. But of course, lines going downhill are equally possible. If we take the line $y = -x$, this has points like $(0, 0)$, $(1, -1)$, $(2, -2)$, and so on. The technical term for going downhill is having a *negative gradient*. This particular example has gradient of -1. How can we see this in the numbers? Well, take the two points on the line, say $(2, -2)$ and $(3, -3)$. The horizontal distance between them is 1, since $3 - 2 = 1$. Notice that we subtracted the x-coordinate of the first point from that of the second point. Doing the same thing for the y-coordinates gives a vertical distance of $(-3) - (-2)$ which, remembering the rules of arithmetic with negative numbers, comes out as -1. This negative distance may seem a strange idea, but all it means is that the graph has *lost* height over that stretch, rather than gaining it. Then applying the usual definition of the gradient, we get a gradient of $\frac{-1}{1} = -1$. If the arithmetic of negative numbers seems confusing, just remember the rule:

> *If the graph goes uphill it has positive gradient; if it goes downhill it has negative gradient; if it is perfectly flat it has a gradient of 0.*

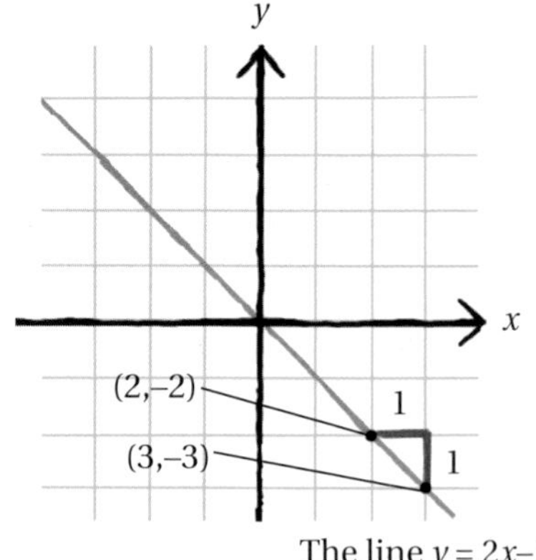

The line $y = 2x-1$

Putting it all together

Let's review the four examples above: the line $y = x$ has gradient 1; the line $y = 2x - 1$ has gradient 2; the line $y = \frac{1}{2}x$ has gradient $\frac{1}{2}$; the line $y = -x$ has gradient -1. There is a pattern here: the gradient always seems to come out as the number by which x is multiplied in the equation. This is not an illusion! So, if a line has equation $y = mx + c$, for two numbers m and c, then its gradient will be m.

This is an extremely useful rule, as it means the gradient can be read straight off the equation, with no intermediate calculations, or pictures, necessary.

The line $y = 5x + 2$ has gradient 5, the line $y = -\frac{3}{20}x - 4$ has gradient $-\frac{3}{20}$, and so on.

A question presents itself: what do the numbers 2 and -4 signify in the previous two equations? More generally, if a line has equation $y = mx + c$, for some numbers m and c, then it has gradient m, but what does the number c represent?

Let's take the line $y = 5x + 2$. What it is the value of y when $x = 0$? The answer is $5 \times 0 + 2 = 2$. Similarly, for the line $y = -\frac{3}{20}x - 4$, when $x = 0$, we get $y = -\frac{3}{20} \times 0 - 4 = -4$. This then is the answer to the question: the number c represents the value of y when $x = 0$.

TRY EXPLORING THIS IN QUIZZES 3 AND 4

In terms of the graph, this denotes the place where the line cuts the vertical y-axis. It is usually referred to as the *y-intercept*.

So we have arrived at a point where we need only two pieces of data to know the equation of a line: its gradient m, and its y-intercept c. Putting these two numbers into the general template $y = mx + c$ gives the line's equation.

This gives a way to tell whether two lines are parallel. Parallel lines have the same gradient (equal values of m) but different y-intercepts (different values of c). So $y = 2x + 1$ and $y = 2x + 2$ are parallel lines, for example, while $y = 3x + 1$ is not parallel to those two.

This is a very useful fact! For example, the rule for converting temperature measured in degrees Celsius into degrees Fahrenheit is to multiply by 1.8 and then add 32. In other words, the equation of this relationship is $y = 1.8x + 32$, where x is a temperature in °C and y is the equivalent temperature in °F. We can now instantly draw a graph of this relationship: a straight line with gradient 1.8 and y-intercept 32.

Sum up *Graphs may seem confusing at first. But remember: the equation of any line is given by inserting just two bits of data into the template $y = mx + c$, namely the gradient (m) and the y-intercept (c).*

Quizzes

1 What are the equations of the straight lines which contain these points?

a (0, 2), (1, 3), (34, 36), (−5, −3)
b (0, 0), (1, 2), (2, 4), (3, 6), (4, 8)
c (0, 0), (1, −1), (2, −2), (3, −3), (4, −4)
d (0, 0), (1, 4), (2, 8), (3, 12), (4, 16)
e (0, 1), (1, 3), (2, 5), (3, 7), (4, 9)

2 Draw a pair of axes, each running from −6 to + 6. Then plot the graphs of these straight lines on it.

a $y = 3x$
b $y = 3x - 1$
c $y = -x$
d $y = -x + 1$
e $y = \frac{1}{3}x + 1$

3 What are the gradients and *y*-intercepts of the lines a–e in quiz 2?

4 Write down the equation of each of these lines.

a The line with gradient 5 and *y*-intercept 4
b The line with gradient −1 and *y*-intercept −1
c The line with gradient $\frac{1}{2}$ and *y*-intercept 2
d The line with gradient −3 and *y*-intercept $\frac{1}{2}$
e The line with gradient 0 and *y*-intercept 8

Statistics

- *Knowing the difference between mean, median and mode*
- *Coping with large piles of data*
- *Understanding quartiles and percentiles*

'There are three kinds of lies: lies, damned lies, and statistics.' So said the novelist Mark Twain, attributing that opinion to British Prime Minister Benjamin Disraeli. It is certainly true that statistics can easily be misunderstood, and abused. Indeed, they often are. But statistical techniques also offer the only tools to navigate through the huge piles of data which emerge the moment we try to extract information in serious quantities from the world.

In this chapter and the next one, we will have a look at some of the contents of a statistician's toolbox, keeping a particular eye out for when they may be 'lying' to us! We'll begin by looking at three different notions of the *average* of a set of data.

The mean, median and mode

The mean is what most people understand by the word *average*. To find the mean of five numbers, we add them all together and then divide by 5. More generally, to find the mean of a group of numbers, we total them all up, and then divide by however many numbers there are.

The mean can be misleading. For example, suppose a group of ten people have a mean income of $100,000 per year. That's ten very well off people, you might think, but it may not be so. It could be the case that one of them is a banker on $1 million per year, and the other nine have no income at all.

This is an example of how *outliers* can distort the statistics of a group. Outliers are individual points which lie a long way away from the general trend of the group (that's the single millionaire in the last example). The mean is particularly susceptible to being skewed by outliers. The next time you hear someone talking about the 'average' of a group (by which they will usually mean the mean), try to imagine how misleading it might be, if the bulk of the total is due to a very few outlying individuals, like the banker in the above example.

Have a practice at Finding the mean in quiz 1.

Another famous example is the paradox of the 'average number of arms'. Suppose there are 1,000 people in a town, of which 999 have two arms, and 1 person has just one arm. The total number of arms is $999 \times 2 + 1$, which is 1,999. So the mean number of arms per person is $1{,}999 \div 1{,}000 = 1.999$. This suggests that almost everyone in town has an 'above average' number of arms!

Despite these warnings, the mean is a very useful statistical tool.

The median is an alternative to the mean. It has two advantages over the mean: it is easier to calculate and is far less skewed by outliers. So what is the median? Suppose I have five children whose heights (arranged in order) are 0.7m, 1.1m, 1.3m, 1.4m and 1.8m. The median, very simply, is the middle value, in this case 1.3m.

Notice that, even if the tallest child is replaced by a 5m-tall giraffe, the median is unchanged: this shows that it is much less sensitive to outliers than the mean. But this also shows that, unlike the mean, the median fails to take all the data into account. (For example, if a team of ten salespeople each get paid $15,000 per year, and the boss wants to know whether she can afford to keep them all on, she needs to know the mean amount of money they bring in, not the median. If the mean is over $15,000, then the team is bringing in more than they cost, but if it is less than $15,000, then the team is losing money.)

Calculating the median of a set of values is easy: first put them in order, and then read off the middle one. The only slight complication arises when there are an even number of values to start with. For example, the weights of four turtles might be: 0.6kg, 4kg, 5kg and 300kg. Here there is no 'middle one' of the four. So instead we take the middle two, 4kg and 5kg, and then take the midpoint of these two values (that is to say, we add the two together and divide by 2). So the median in this case is 4.5kg.

NOW TRY QUIZ 2.

The mode While the mean and median are the most useful averages, it is worth mentioning this third type of average. The mode of a set of numbers is the easiest to calculate: it is simply the number that occurs most often. Suppose I survey the houses on my road, asking how many people live in each house. Suppose the answers are that: 1 house is empty, 19 have 1 person living there, 23 have 2 people living there, 15 have 3 people, and 9 have 4 people. The mode is just the commonest answer, in this case 2.

One advantage of the mode is that it works just as well with data that are not numerical. If three people in a house have red hair, one has black hair and another has blonde hair, then it makes perfectly good sense to say that the *modal* hair colour in that household is red.

The mode can be useful, but it is limited. For example, if I measure the heights of the people in my family to the nearest millimetre, it is very unlikely that two answers will be the same. So there is no meaningful mode. (Even if there are identical twins in my family who do have the same height, that answer does not carry much information.) The mode is only relevant when the number of possible responses is limited, which is not the case with height.

It is the case in elections, though, which is why the winning candidate is precisely the *modal* candidate, that is to say the one who achieves the largest number of votes.

Mountains of data!

We have learned the basic techniques for dealing with averages. There is just one problem: all the examples we have seen so far have involved only small sets of data. Statistical techniques really come into their own when large quantities of data need to be interpreted. So let's see what happens! The mathematics will be exactly the same as above, but the way we write and think about it will have to be adjusted.

If we have a hundred or a thousand data points, instead of five or six, the first change is that, instead of writing the data in a huge list, we will make life easier by entering it in a table.

Suppose we survey the residents of an island, asking how many times they have travelled abroad in the last year:

NUMBER OF TRIPS ABROAD	NUMBER OF PEOPLE
0	102
1	745
2	591
3	373
4	122
5	46
Total	**1,979**

The table tells us that 102 people travelled abroad 0 times in the last year, 745 had 1 trip away, and so on.

With this new presentation of the data, we want to be able to calculate various statistics. One statistic can just be read straight off: the mode. The commonest response to the survey is '1', so that is the mode.

What of the median? Listing the residents of the island in order of their number of trips, we want to know where the middle one is. Well, we know that the total number of people is 1,979, so the middle person is number 990. (This is calculated by adding 1 and then dividing by 2.) Now there is a slight obstacle, in that the table does not immediately tell us where person number 990 sits in the table.

The best way to approach this is to add a new column. First we will rename the 'number of people column' the *frequency*, as it tells us how frequent each response to the survey is. Then we add a new column to the right, called the *cumulative frequency*. 'Cumulative' means 'adding up as we go along' (as in 'accumulate'). That is exactly what the new column contains: the frequencies added together as we move down the column.

NUMBER OF TRIPS ABROAD	FREQUENCY	CUMULATIVE FREQUENCY
0	102	102
1	745	847
2	591	1,438
3	373	1,811
4	122	1,933
5	46	1,979

The interpretation of the new column is this: 102 people have had 0 trips, $102 + 745 = 847$ have had 0 or 1 trips, then 1,438 have had at most 2 trips (0, 1 or 2 trips), and so on. The useful thing about this column is that we can now immediately locate the median, that is to say the 990th person.

If we pretend that each person has a number, then people 1–102 have each travelled 0 times, people numbered 103–847 have each travelled once, and people 848–1,438 have each travelled twice. It is this category that contains the 990th person. So the median number of trips is 2.

Quartiles and the interquartile range

There is other valuable information that can easily be extracted from a cumulative frequency table.

TRY THIS LINE OF THINKING IN QUIZ 4.

In the example above, the cumulative frequency table arranges the people into increasing order of the number of trips they have taken. With this done, the median is the value at the half-way point, which we identified as 2. We can equally well ask about the values at the quarter-way point, orthe three-quarter-way point. These are known as the first and third *quartiles.* (The second quartile is the median.)

To calculate the first quartile, the only tricky part is to work out where the quarter-way mark is. Once that is done, the answer can be read straight off the cumulative frequency table.

In the example above, there are 1,979 people, and the quarter-way point is found by adding 1 and then dividing by 4 (or equivalently multiplying by $\frac{1}{4}$), which gives the 495th person. The cumulative frequency column tells us that this sits in the row corresponding to 1 trip abroad, so the first quartile is 1.

The median, mode, quartiles and percentiles can all be read off a cumulative frequency table almost immediately.

Similarly the three-quarter mark is found by adding 1 and then multiplying by $\frac{3}{4}$, which gives us the 1,485th person. This sits in the row corresponding to 3 trips, so the third quartile is 3.

The mean, median and mode are all measures of where the centre of a set of data is. Another fact we would like to know is how *spread out* it is. A common measure of this is the interquartile range:

> *The* interquartile *range is the difference between the first and third quartiles.*

So, in the example above, the interquartile range is $3 - 1 = 2$.

Percentiles: sifting more finely

The same line of thinking which gave us quartiles also produces *percentiles*, which are a finer sifting of the data. Surprisingly the exact definition of a *percentile* is not fully agreed among statisticians, but in most cases the answers come out the same. We'll take the very simplest option.

The quartiles divided the data into four parts. Similarly, we can divide it into one hundred percentiles. To calculate the 95th percentile (for example), take the total number of data points, which is 1,978 in the above example, and then multiply that by the decimal which corresponds to the percentile, which is 0.95. So we get $1{,}978 \times 0.95 = 1{,}879.1$. Now we want to find that number in the cumulative frequency table, and we spot that it is in the row corresponding to 4 trips. So in this case the 95th percentile is 4. Similarly, the 99th percentile is at the point $1{,}978 \times 0.99 = 1{,}958.22$, which is in the row of the table corresponding to 5 trips abroad, giving an answer of 5.

HAVE A GO AT CALCULATING PERCENTILES IN QUIZ 5.

Sum up *From the mean to the 99th percentile, the statistician's toolbox contains many useful devices for making sense of piles of data!*

Quizzes

1 Find the mean of each of these sets of data.

a The weights of five people (to the nearest kg): 25kg, 42kg, 60kg, 34kg, 45kg

b The lengths of three snakes (to the nearest 0.1m): 1.7m, 0.4m, 0.9m

c The number of CDs owned by four people (exact answers): 0, 28, 12, 143

d The loudness of six dogs' barks (to the nearest decibel): 21db, 46db, 19db, 35db, 51db, 27db

e The number of televisions in each house in street: 2 houses have no TV, 16 houses have 1 TV, 8 have 2, and 4 have 3.

2 For each set of data in quiz 1, find the median.

3 Write a cumulative frequency column for this table. Then find the median and mode.

The number of pets per household in a village:

NUMBER OF PETS	NUMBER OF HOUSEHOLDS
0	497
1	695
2	558
3	380
4	224
5	72

4 Find the quartiles, the interquartile range and the 99th percentile for the data in quiz 4.

Probability

- *Knowing how to use numbers to measure likelihood*
- *Analysing combinations of events*
- *Understanding how different events can affect each other*

There are many aspects of the world that can be measured with numbers. This is what makes mathematics is so endlessly fascinating! But this is not limited to things that we can weigh or measure. Some applications of numbers are subtler, and more indirect. One important area is the study of* probability*, where we use numbers to represent the likelihood of certain events taking place.

When we talk about an event being 'likely' to happen, or 'certain' to happen, we are using the language of probability. In the study of probability, we assign a number to this likelihood to quantify the chance of the event happening. We use only the numbers between 0 and 1: an impossible event has probability 0, while a certainty has a probability of 1. Everything else falls somewhere in between. For example, if I toss a coin, then the probability of it landing on heads is $\frac{1}{2}$ (one chance in two), so long as the coin is fair.

GOLDEN RULE

The numbers between 0 and 1 measure how probable an event is, with impossible corresponding to 0 and certain to 1.

Biased coins will have other probabilities. To take an extreme example, a double-headed coin (a coin with a head on each side) will have probability 1 of landing on heads (it is certain to happen). For the rest of this chapter we will make a standing assumption that all the coins (and dice and decks of cards) we meet are fair.

At one end of the scale, unlikely events have very small probabilities, meaning numbers close to 0. The chance of your ticket winning the UK National Lottery or the Washington State Lottery are each $\frac{1}{13{,}983{,}816}$ (around 0.00000007). Events which are completely impossible have a probability of 0. (As lotteries like to advertise: if you don't have a ticket, your chance of winning is absolutely nil!)

At the other end of the scale, very likely events have high probabilities, meaning numbers close to 1. The probability that the sun will rise tomorrow is very close indeed to 1, something like 0.9999. . .999. (I wouldn't want to guess the number of 9s, but in the 18th century the naturalist George-Louis Leclerc made a serious attempt to estimate it!)

If I am asked to give a rough and ready estimate of how likely it is to rain tomorrow, I might start by reasoning that, at this time of year, it typically rains in my town around one day in two. This would put the chance of it raining tomorrow at around 0.5. If it has been raining across the entire region for several days and shows no signs of clearing up, I might increase that estimate to, say, 0.8.

TRY ESTIMATING SOME PROBABILITIES IN QUIZ 1.

Counting successes

It is all very well estimating the probabilities of events according to how likely they *seem*. But how can we work out exact answers? One basic technique amounts to counting the outcomes of an experiment.

When we roll a standard die, there are six possible outcomes (1, 2, 3, 4, 5, 6). Suppose I want to know the probability of rolling a 5. Just one of the six outcomes counts as a 'success', which gives us our answer: a probability of $\frac{1}{6}$. The rule here is that the *total* number of possible outcomes goes on the bottom of the fraction, and the number of 'successful' outcomes goes on the top:

$$\text{Probability of an event} = \frac{\text{Number of successful outcomes}}{\text{Total number of outcomes}}$$

This is the basic idea. But as usual there is some fine print to take account of! If we go back to the question of whether or not the sun will rise tomorrow, then there are two possible outcomes: either it will or it won't. Of these, just one (sunrise), is classed as a 'success'. So, according the rule above, the answer should be $\frac{1}{2}$.

The trouble with this is obvious! It is just nonsense. Sunrise is a near certainty, and so should have a probability very close to 1.

So what has gone wrong? Well, when counting up successes and outcomes, there is an additional rule: that *all the possible outcomes must be equally likely*. This is what fails in the case of the sun. So, for the formula to work, the dice and coins used must be fair.

IT'S TIME FOR QUIZ 2.

Combining events: 'and'

What is the point of assigning numbers to the probabilities of events? It is not just because mathematicians are fixated with measuring everything numerically. One benefit is that different ways of combining events correspond very neatly to various arithmetical tricks with their probabilities. There are two principal cases of this, which are described by the two English words 'and' and 'or'.

Let us take 'and' first. Suppose I roll a die and flip a coin. What is the probability that I will roll a 6 *and* flip a head? We know that the probability of rolling a 6 is $\frac{1}{6}$, and the probability of getting a head is $\frac{1}{2}$. How do we mix these numbers, to get the probability for the combined event of a head *and* a 6. The answer is to multiply. So the probability we want is $\frac{1}{6} \times \frac{1}{2}$, which comes out as $\frac{1}{12}$.

The general rule here is that 'and' in a combined event means 'multiply' the probabilities. But we cannot just apply this rule blindly; again there is some fine print to take into account. What, for example, is the probability that, when I flip a coin once, I get both a head and a tail? The answer should be 0, since that event is completely impossible. But if we apply the rule above, without thinking about what it means, we get an answer of $\frac{1}{2} \times \frac{1}{2} = \frac{1}{4}$.

What is the caveat we need to eliminate nonsense like this, and leave us with a rule that makes sense? The answer is that the two events whose probabilities we are multiplying must not affect each other. In technical terms, they must be *independent*. The first example passes this test: when I roll a die and toss a coin, whether or not I get a head has no impact on whether or not I get a 6. But, in the second example, with just one coin, whether or not I get a head makes a huge difference to the likelihood of my getting a tail. (In fact the one determines the other entirely.)

So we can express the rule more accurately: when two events are independent, 'and' means 'multiply'.

IT'S TIME TO HAVE A GO AT QUIZZES 3 AND 4.

Combining events: or'

Let's move on to the other principal way that two events can be combined: 'or'. When I roll a die, I might be interested in the probability of my getting a 5 *or* a 6. Here we can move directly to the method of counting up successes and outcomes, which will quickly give us the answer: $\frac{2}{6}$ (which can be simplified to $\frac{1}{3}$: see *Fractions*). But it is useful to think about how this answer is related to the individual probabilities of the two separate events: getting a 5 or getting a 6. Each of these has probability $\frac{1}{6}$. The probability of the combined event, a 5 *or* a 6, comes from adding these two together.

So the general rule is, when finding the probability of a combined event: 'or' means 'add'.

Let's have another example: Suppose I pick a card from a deck. The probability of getting a heart is $\frac{1}{4}$. The probability of getting the queen of spades is $\frac{1}{52}$. So what is the probability of getting a heart or the queen of spades? Well we can apply the simple rule – 'or' means 'add' – to get an answer of $\frac{1}{4} + \frac{1}{52}$ which comes out as $\frac{7}{26}$ once the fractions have been added and simplifed. (Try working that through yourself!)

As ever, though, caution is needed because this rule also comes with some fine print. Here is why: Suppose I flip two coins. What is the probability that I will get two heads? If I unthinkingly apply this rule, I would reason as follows: The probability that I get a head on the first coin is $\frac{1}{2}$. The probability that I get a head on the second coin is $\frac{1}{2}$ too. 'Or' means 'add', so the probability that I get a head on the first coin *or* the second coin is $\frac{1}{2} + \frac{1}{2} = 1$. This suggests that it is a certainty. But of course this is nonsense: it is perfectly possible that I will get two tails.

The fine print in this case is that you can only add together the probabilities of two events when *they cannot both occur*. When I roll a die, I cannot get both a 5 and a 6. So it is safe to add together those probabilities. But I can get heads on two coins, so I am not allowed just to add together those probabilities. In the jargon, the two events must be *mutually exclusive*. This means that if one happens, then the other doesn't.

If I roll one die, the two outcomes (a 5 and a 6) are mutually exclusive. But if I flip two coins, the two outcomes (a head and a head) are not mutually exclusive.

HAVE A GO AT QUIZ 5

Now we can express the rule more accurately: when two events are mutually exclusive, 'or' means 'add'.

Sum up *Whenever you think something is 'impossible', 'unlikely' or 'certain', you are using the language of probability. It has techniques to assess the likelihood of different events happening – a valuable prize in this uncertain world!*

Quizzes

1 *Guess approximate probabilities for the following events (answers may vary from person to person!).*

a *The next person you meet will be male.*
b *An asteroid will hit your house tomorrow.*
c *If you turn on the TV, the first person you see will be wearing glasses.*
d *If you pick a word on this page at random, it will have an 'e' in it.*
e *Your favourite sports team will win their next match.*

2 *Calculate these probabilities by adding up the total number of successes and outcomes.*

a *You pick a playing card from an ordinary deck. What is the chance of getting an ace?*
b *You roll an ordinary die. What is the chance of getting an even number?*
c *You roll a 12-sided die. What is the chance of getting an 8 or higher?*
d *You pick a card from an ordinary deck. What is the chance of getting a spade?*
e *You roll a 20-sided die. What is the chance of getting a prime number?*

3 Are these pairs of events independent?

a You toss a 10c coin and a 25c coin and get a head on the 10c and a tail on the 25c coin.

b You roll a die and pick a card from a deck. You get an even number on the die and a king.

c You pick a card from a deck, replace it, shuffle, and pick another card. Both times you get an ace.

d You pick a card from a deck, don't replace it, and then pick another. Both times you get an ace.

e You pick a single card from a deck and get a black card and an ace.

4 In quiz 3 above, where the pairs of events are independent, calculate the combined probability of both occurring.

5 Are these pairs of events mutually exclusive? Where the answer is yes, calculate the probability of one or the other happening.

a You pick a card from a deck and get the queen of spades or a heart.

b You roll a die and get an odd number or a 6.

c You toss a 10c coin and a 25c coin and get a head on the 10c coin or a head on the 25c coin.

d You roll two ordinary dice, and their total is 2 or their total is 12.

e You pick a card from the deck, replace it, shuffle and pick again. You get the ace of spades twice, or the queen of hearts twice.

Charts

- *Understanding how to interpret pie charts and bar charts*
- *Representing proportion graphically*
- *Knowing how to convert raw data to charts*

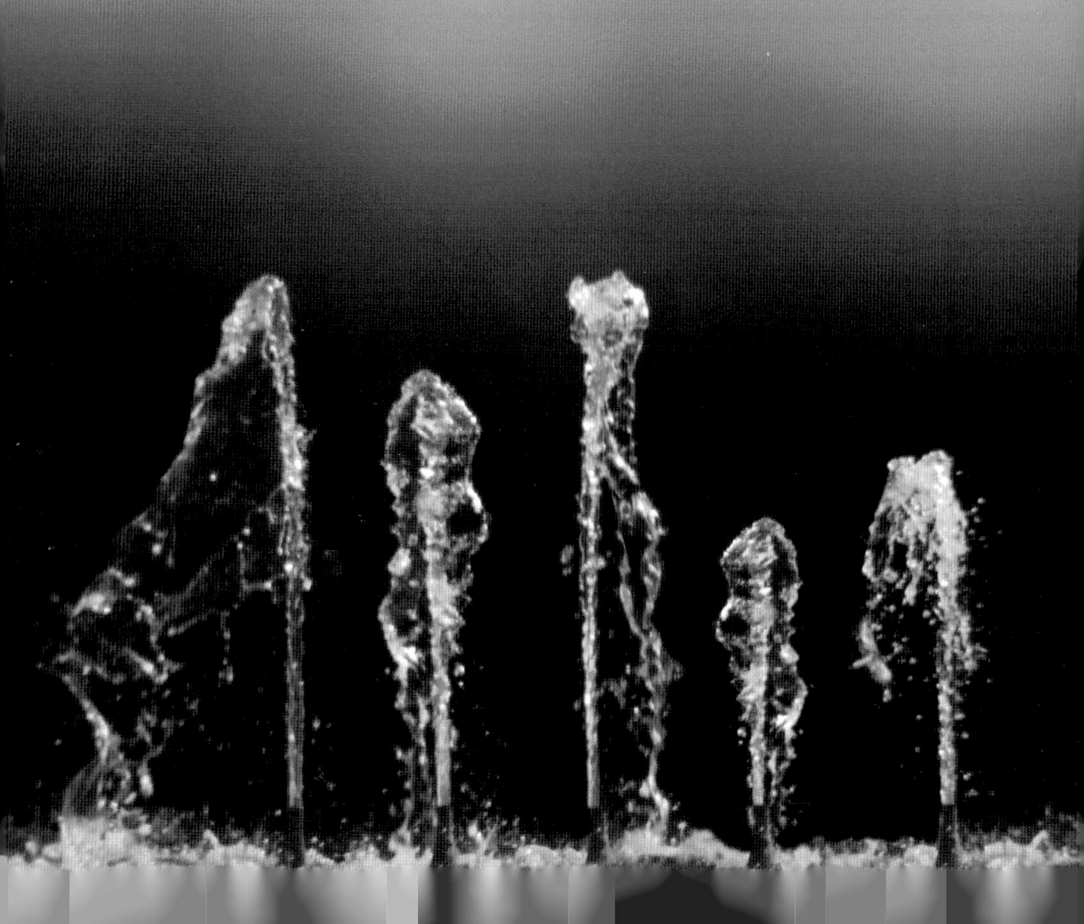

Every day, in every newspaper, magazine and current-affairs website, you will find a wealth of statistics. Often, though, these numbers are not displayed in tables or lists but are incorporated into diagrams of the statistics. There are several different types of diagram or chart, and most are easy to understand visually. Indeed, this is why they are used!

In this chapter, we will have a look at these charts in more depth; you will see how to interpret them and how to understand the rules for producing charts yourself. Most spreadsheet programs have tools for creating such charts, and if you want nice-looking charts for a presentation, that is the best way to proceed. You can think of this chapter as a behind-the-scenes glimpse of what these programs do.

Pie charts

The first type of diagram we will look at is the *pie chart*. The name is 'pie' as in something delicious baked in the oven, rather than the number π that we met in *Circles*. But, as it happens, both are relevant, since a pie chart is essentially a circle, divided up into differently coloured slices. The idea is that the size of each slice corresponds to some proportion of the whole.

Let's take an example. In my city, one third of the people have blonde hair, one third brown and one third black. To represent this information in a pie chart is straightforward: first draw a circle, and then divide it into three slices. One slice represents the people with blonde hair, another those with brown hair, and the third those with black hair. Crucially, in this case, the three slices must be the same size, because the three sections of the population are the same size.

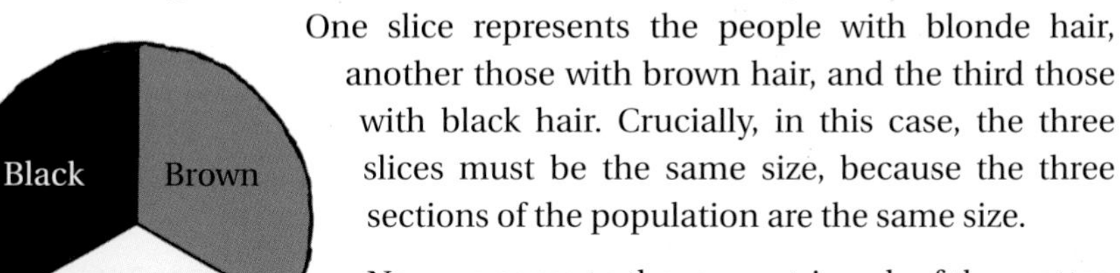

Now we come to the geometric nub of the matter. How do we divide a circle into thirds? It is easy enough to do it approximately by eye, but we want to do it exactly. The slices will all meet at the centre of the circle. The key to the matter is the *angle* of each slice. Because the three slices are all intended to have equal size, the three angles must be equal too. What is more, in the terminology of the chapter *Angles*, these

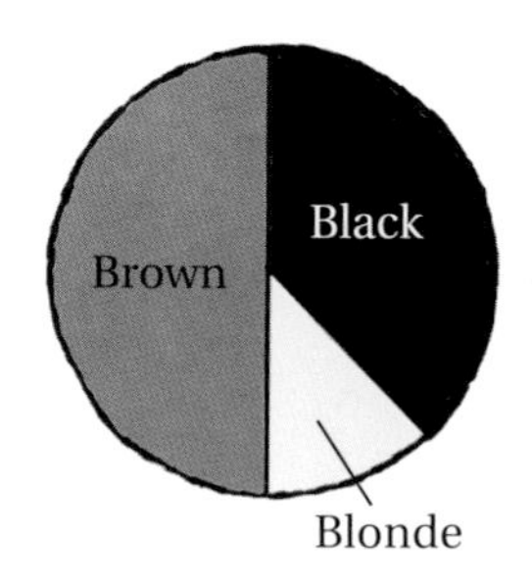

form *angles at a point*, which means that they must add up to 360°. So we are looking for three equal angles which add to 360°. It is obvious, I hope, that each one must be $360 \div 3 = 120°$. Once the angles have been established, it is just a matter of drawing the chart, using a protractor.

Of course, things are trickier when the proportions are not all the same. In the next town along, half the people have brown hair, three eighths have black hair and just one eighth have blonde hair. How can this be represented as a pie chart? The clue is in the proportions: one half, three eighths and one eighth. All we need to do is split up the angle at the centre of the pie according to these proportions. This amounts to multiplying 360° by each of the proportions in turn. To start with, $\frac{1}{2} \times 360 = 180°$. This represents the largest slice: the half of people who have brown hair. (It should not be a surprise that 180° looks like a straight line.) Next, $\frac{3}{8} \times 360 = 135°$. (You can calculate this by multiplying 360 by 3 and then dividing by 8.) Finally, $\frac{1}{8} \times 360 = 45°$. With these angles, it is now easy enough to draw the chart.

HAVE A GO AT THIS YOURSELF IN QUIZZES 1 AND 2.

From raw data to proportions

If we want to draw a pie chart, we need to be able to calculate proportions from the raw statistics, rather than having the proportions given to us. Usually, things will not come out as neatly as $\frac{1}{2}$ or a $\frac{1}{4}$! But precisely the same

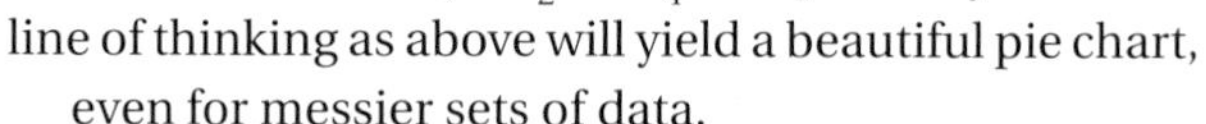
line of thinking as above will yield a beautiful pie chart, even for messier sets of data.

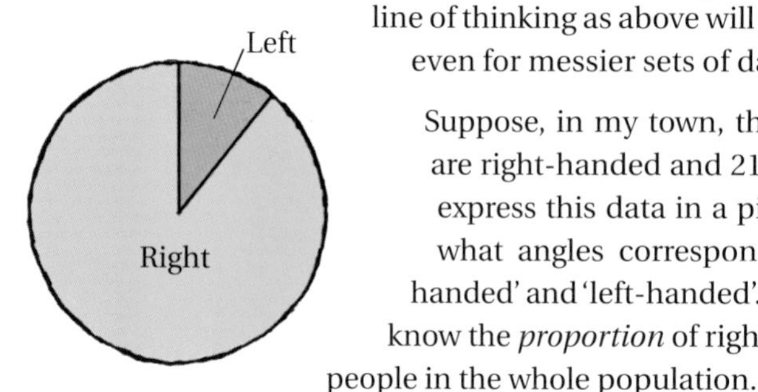

Suppose, in my town, there are 1,794 people who are right-handed and 215 who are left-handed. To express this data in a pie chart, we need to know what angles correspond to the slices for 'right-handed' and 'left-handed'. To do that, we first need to know the *proportion* of right-handed and left-handed people in the whole population.

The total number of people is $1794 + 215 = 2009$. So the proportion of right-handed people is $\frac{1794}{2009}$, and that of left-handed people is $\frac{215}{2009}$. We could convert these to decimals or percentages, but there is no need.

Instead, we can now work out the angles for the pie chart. The angle corresponding to right–handed people is now $\frac{1794}{2009} \times 360°$, which we calculate by multiplying 360 by 1,794, and then dividing by 2,009, to give an answer of 321.5° (to one decimal place). Similarly $\frac{215}{2009} \times 360° = 38.5°$ (to one decimal place). These are the angles needed for the pie chart, which can then be drawn easily.

TRY THIS YOURSELF IN QUIZ 3.

Bar charts

Even more familiar than pie charts are *bar charts*. The starting point this time is a pair of axes, similar to those we used in *Coordinates*. The vertical axis has the scale on it. So if we are measuring the populations of countries, then the vertical axis might be labelled with numbers such as 10 million, 20 million, 30 million, and so on.

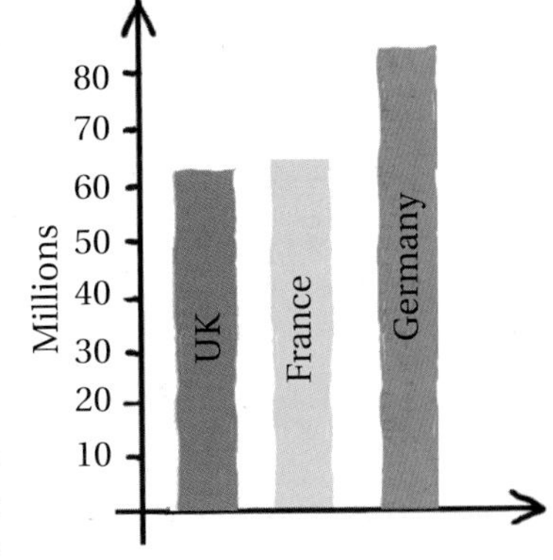

The horizontal axis is the ground from which bars grow. The first bar, for example, might represent the UK: the *height* of the bar carries the data. In this case, it represents the population of the UK, around 63 million. Then the next bar represents the next country, and so on. It is good practice to keep the bars equally spaced.

(Sometimes you see bar charts with the axes switched round, so that the bars extend from left to right, as if running a race, rather than growing from the ground, like buildings.)

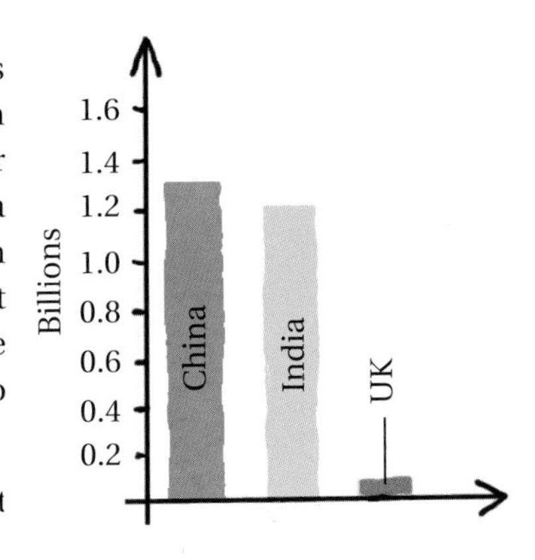

The idea of a bar chart is simple enough: this is why they are so commonly used. The main element of skill is in choosing the right scale for the data. For example, if I wanted to include China (population 1.3 billion) and India (1.2 billion) in the above chart, then the scale I used above is not suitable, as the bars for China and India would be too long to fit on the page. It would be better to have a scale in billions this time.

Of course this makes the UK bar very small, but that's unavoidable.

On the other hand, if I was comparing the populations of various small islands such as Fair Isle (population 72) and Bressay (population 390), then a scale which increased in hundreds would be better.

So, the first step to creating a bar chart is to look at the numbers involved, and choose a sensible scale. Pick a maximum number: ideally a nice round number, which is slightly more than the biggest number you need to represent on the chart. I might pick 500 for the islands example above. After that, all that remains is to draw the axes, write in the scale on the vertical axis (making sure that the numbers increase in equal steps, not in uneven jumps), then draw in bars with the correct heights, and label the bars so we know what's what.

Segmented bar charts

There are various ways in which bar charts can be spiced up to represent subtler forms of data. One such is the *segmented* bar chart.

GOT IT?
TRY QUIZ 4.

What is the purpose of it? Well, in some ways it combines the strengths of a pie chart and a bar chart. Suppose we want to make a chart representing the population of a certain country, and how this has evolved over the years 2002–2011. A bar chart seems a good choice here, with ten bars representing the ten years, and the heights of the columns representing the population that year.

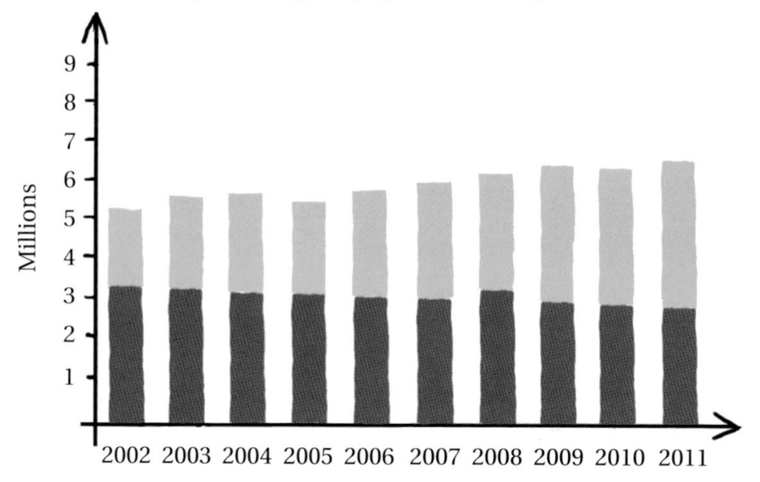

We might also be interested in what *proportion* of population is in full-time work, for example. If we were *only* interested in those proportions, then a pie chart would be a better choice. We could have a sequence of ten pie charts, one for each year. Each pie chart would be cut into two slices: one representing people who are in full-time work, and the other representing those who are not, making it easy to compare the proportions. Notice, though, that these pie charts contain no information about actual numbers of people. As our golden rule tells us, they are solely about *proportions* of the population, not about the *size* of the population.

GOLDEN RULE

Some types of graph (e.g. bar charts) represent the actual sizes of populations. Others (e.g. pie charts) represents relative proportions within a population. Be sure you know the difference!

We might want both of these types of data in our chart: absolute numbers and proportions. In this case, the best option is to take the bar chart and slightly amend it. Each bar represents the entire population in one year. The idea is to split it into two segments: one representing the people in full-time work, and the other those who are not. Now we can easily compare the size of the full population from year to year, as well as the numbers of people in full-time work.

IT'S TIME FOR QUIZ 5

Sum up *There are many different ways to represent data: bar charts, pie charts and segmented bar charts are just a few. All of them are easy to understand visually – after all, that is the point!*

Quizzes

1 Calculate the angles needed for these pie chart.

a In a block of flats, a quarter of the flats have one resident, half have two, and a quarter have three or more.

b On a menu, two thirds of the dishes are vegetarian, one sixth contain meat, and one sixth contain fish.

c One day, a TV channel dedicates one fifth of the time to news, three fifths of the time to drama, one tenth of the time to music, and one tenth to advertising.

d In a bookshop, five eighths of the books are fiction, one eighth are reference, one eighth are biography and one eighth are assorted non-fiction.

e In my house, seven tenths of the wall-space is painted, one fifth is wallpapered, and one tenth is tiled.

2 Draw a pie chart for each of the scenarios in question 1.

3 Calculate the proportions, and angles, needed for the pie charts in these situations.

a A room contains 10 men and 12 women.

b A radio station dedicates 1 hour a day to news, 1 hour to advertising and the rest to music.

c One month, a cinema shows 8 thrillers, 15 comedies, 2 horror films and 5 children's films.

d A man's CD collection contains 124 rock albums, 17 jazz albums, 36 classical albums and 3 spoken word albums

e A page of text contains 104 nouns, 76 verbs, 25 adjectives and 18 adverbs.

4 Draw a bar charts to represent each of these situations.

a In a village, 34 households have only cats, 75 have only dogs, 17 have both cats and dogs, while 88 have neither cats nor dogs.

b A class of students take a test. Their marks out of 5 are:

SCORE	0	1	2	3	4	5
NUMBER OF STUDENTS	2	5	7	13	11	8

c An orchestra contains 27 string players, 14 wind players, 8 brass players and 3 percussionists.

d A jungle contains 14.5 million herbivorous animals, 1.6 carnivorous animals and 10.1 million trees.

e A shop tracks its number of customers: in 2009 there were 10,225, in 2010 there were 12,987, in 2011 there were 15,011 and in 2012 there were 14,991.

5 Draw segmented bar charts for these sets of data. (Notice that sometimes the first type of data includes the second, and sometimes the two are separate.)

a The number of people in a village:

YEAR	1970	1980	1990	2000	2010
ADULTS	167	179	184	199	171
CHILDREN	14	27	23	32	26

b The number of crimes reported to police:

MONTH	Jan	Feb	Mar	Apr	May
TOTAL CRIMES	1047	979	891	909	870
VIOLENT CRIMES	377	280	314	390	451

Answers to selected quizzes

The language of mathematics page 4: **Quiz 1, a** $10 + 11 = 21$ (true), **b** $2 \times 2 = 2 + 2$ (true), **c** $5 - 4 = 2 \div 2$ (true), **d** $5 \div 2 \geqslant 3$ (false), **e** $5 \times 4 < 3 \times 7$ (true). **Quiz 2, a** $(1 + 2) + 3 = 6$ & $1 + (2 + 3) = 6$, **b** $(4 + 6) \div 2 = 5$ & $4 + (6 \div 2) = 7$, **c** $(2 \times 3) \times 4 = 24$ & $2 \times (3 \times 4) = 24$, **d** $(20 - 6) \times 3 = 42$ & $20 - (6 \times 3) = 2$, **e** $(2 \times 3) + (4 \times 5) = 26$ & $2 \times (3 + 4) \times 5 = 70$. **Quiz 3, a** Both correct, **b** $4 + (6 \div 2) = 7$, **c** Both correct, **d** $20 - (6 \times 3) = 2$, **e** $(2 \times 3) + (4 \times 5) = 26$. **Quiz 4,** $-$ & $\div$

Addition page 11: **Quiz 1, a** 11, **b** 13, **c** 18, **d** 12, **e** 21. **Quiz 2, a** 70, **b** 7000, **c** 1100, **d** 11,000, **e** 120,000. **Quiz 3, a** 78, **b** 99, **c** 698, **d** 785, **e** 9775. **Quiz 4, a** 41, **b** 74, **c** 161, **d** 471, **e** 401. **Quiz 5, a** 59, **b** 87, **c** 101, **d** 93, **e** 161. **Quiz 6, a** 83, **b** 89, **c** 86, **d** 97, **e** 144.

Subtraction page 18: **Quiz 1, a** 3, **b** 4, **c** 7, **d** 8, **e** 8. **Quiz 2, a** 21, **b** 28, **c** 430, **d** 2838, **e** 2140. **Quiz 3, a** 54, **b** 17, **c** 34, **d** 19, **e** 99. **Quiz 4, a** 37, **b** 61, **c** 22, **d** 49, **e** 58. **Quiz 5, a** 38, **b** 55, **c** 26, **d** 38, **e** 42.

Multiplication page 24: **Quiz 1, a** 16, **b** 30, **c** 54, **d** 49, **e** 56. **Quiz 2, a** 68, **b** 88, **c** 93, **d** 128, **e** 205. **Quiz 3, a** 880, **b** 690, **c** 480, **d** 1890, **e** 3550. **Quiz 4, a** 714, **b** 1530, **c** 2790, **d** 8733, **e** 54,864. **Quiz 5, a** 912, **b** 2074, **c** 1653, **d** 11,096, **e** 26,292.

Division page 35: **Quiz 1, a** 3, **b** 6, **c** 9, **d** 8, **e** 7. **Quiz 2, a** 2 r 3, **b** 2 r 4, **c** 3 r 3, **d** 9 r 3, **e** 7 r 3. **Quiz 3, a** 12, **b** 22, **c** 31, **d** 62, **e** 1061. **Quiz 4, a** 432, **b** 110, **c** 301, **d** 4241, **e** 3012. **Quiz 5, a** 121, **b** 142, **c** 131, **d** 142, **e** 124. **Quiz 6, a** 21, **b** 18, **c** 31, **d** 105, **e** 61.

Primes, factors and multiples page 44: **Quiz 1, a** 3×5, **b** $2 \times 3 \times 3$, **c** 3×7, **d** $2 \times 2 \times 2 \times 3$, **e** $2 \times 2 \times 2 \times 2 \times 2$. **Quiz 2, a** True & false, **b** False & false, **c** False & false, **d** True & true, **e** True & true. **Quiz 3, a** Divisible by 2, 4, & 8, **b** Divisible by 2, 3, 6, & 7, **c** Divisible by 3 & 5, **d** Divisible by 2, 4, 8, & 11, **e** Divisible by 2 & 4. **Quiz 4, a** $2 \times 3 \times 5$, **b** $2 \times 3 \times 5 \times 7$, **c** $2 \times 2 \times 3 \times 3 \times 3$, **d** $3 \times 3 \times 3 \times 7$, **e** $3 \times 7 \times 7 \times 11$. **Quiz 5, a** $3 + 7$ or $5 + 5$, **b** $5 + 7$, **c** $3 + 11$ or $7 + 7$, **d** $3 + 13$ or $5 + 11$, **e** $5 + 13$ or $7 + 11$.

Negative numbers and the number line p 50: **Quiz 2, a** 15 & 1, **b** 6 & 0, **c** 9 & −3, **d** −1 & −9, **e** 1 & −5. **Quiz 3, a** 1 & 9, **b** −1 & 5, **c** −5 & 5, **d** −6 & −2, **e** −8 & 2. **Quiz 4, a** −6 & 6, **b** −20 & 20, **c** −21 & 21, **d** −32 & 32, **e** −100 & 100. **Quiz 5, a** 4 & −4, **b** −3 & 3, **c** 4 & −4, **d** −11 & 11, **e** −3 & 3.

Decimals page 57: **Quiz 1, a** 5.5, **b** 13.1, **c** 15.50, **d** 10.11, **e** 0.01111. **Quiz 2, a** 4.3, **b** 4.12, **c** 7.9, **d** 4.19, **e** 9.83. **Quiz 3, a** 0.8, **b** 0.08, **c** 3.5, **d** 0.35, **e** 0.035. **Quiz 4, a** 3.91, **b** 32.24, **c** 9.86, **d** 174.76, **e** 45.188. **Quiz 5, a** 5.3, **b** 0.16, **c** 0.2, **d** 10.0, **e** 0.720.

Fractions page 66: **Quiz 1, a** $\frac{9}{12}$, **b** $\frac{6}{8}$, **c** $\frac{12}{16}$, **d** $\frac{15}{20}$, **e** $\frac{75}{100}$. **Quiz 2, a** $\frac{1}{2}$, **b** $\frac{2}{3}$, **c** $\frac{3}{7}$, **d** $\frac{5}{6}$, **e** $\frac{4}{5}$. **Quiz 3, a** $1\frac{3}{4}$, **b** $3\frac{1}{3}$, **c** $2\frac{3}{4}$, **d** $4\frac{1}{2}$, **e** $3\frac{2}{7}$. **Quiz 4, a** $\frac{9}{10}$, **b** $\frac{3}{5}$, **c** $\frac{19}{20}$, **d** $\frac{5}{8}$, **e** $\frac{7}{8}$. **Quiz 5, a** 0.8, **b** $0.\dot{6}$, **c** 0.1875, **d** $0.\overline{285714}$, **e** $0.\dot{5}$. **Quiz 6,** All numbers except multiples of 2 & 5.

Arithmetic with fractions p73: **Quiz 1, a** $\frac{3}{15}$ or $\frac{1}{5}$, **b** $\frac{29}{100}$, **c** $\frac{25}{29}$, **d** $\frac{27}{33}$, **e** $\frac{28}{999}$. **Quiz 2, a** $\frac{2}{9}$, **b** $\frac{11}{12}$, **c** $\frac{7}{10}$, **d** $\frac{73}{100}$, **e** $\frac{24}{25}$. **Quiz 3, a** $\frac{1}{15}$, **b** $\frac{7}{12}$, **c** $\frac{11}{20}$, **d** $\frac{11}{12}$, **e** $\frac{19}{30}$. **Quiz 4, a** $\frac{1}{4}$, **b** $\frac{2}{5}$, **c** $\frac{1}{2}$, **d** $\frac{1}{6}$, **e** $\frac{1}{60}$. **Quiz 5, a** $\frac{3}{4}$, **b** $\frac{1}{2}$, **c** $\frac{7}{16}$, **d** $\frac{7}{4}$, **e** $\frac{1}{18}$.

Powers page 80: **Quiz 1, a** 27, **b** 36, **c** 125, **d** 81, **e** 216. **Quiz 2, a** 3125, **b** 46,656, **c** 20,736, **d** 15,625, **e** 5,584,059,449. **Quiz 4, a** $\frac{1}{10{,}000}$, **b** $\frac{1}{128}$, **c** $\frac{16}{81}$, **d** $\frac{1}{1{,}000{,}000}$, **e** $\frac{81}{256}$. **Quiz 5, a** 2^{22}, **b** 2^{15}, **c** 15^{5}, **d** 13^{26}, **e** 100^{19}.

The power of 10 p 86: **Quiz 1, a** 7,000,000, **b** 8,000,000,000, **c** 9,000,000,000,000, **d** 10,000,000,000,000,000, **e** 11,000,000,000,000,000,000. **Quiz 2, a** 18 kilometres, **b** 37 megapixels, **c** 3 nanograms, **d** 900 kilonewtons or 0.9 meganewtons, **e** 8 gigabytes. **Quiz 3, a** one thousandth, **b** one hundredth, **c** one hundred thousandth, **d** one ten millionth, **e** one trillionth. **Quiz 4, a** 600,000, **b** 21000, **c** 0.00000879, **d** 0.001332, **e** 67,100,000,000. **Quiz 5, a** 8×10^{5}, **b** 5.6×10^{4}, **c** 6.2×10^{-4}, **d** 9.87×10^{8}, **e** 1.11×10^{-9}.

Roots and logs p 95: **Quiz 1, a** 2, **b** 10, **c** 8, **d** 7, **e** 12. **Quiz 2, a** 3.16, **b** 2.15, **c** 1.78, **d** 2.24, **e** 1.16. **Quiz 3, a** 11, **b** 2, **c** 8, **d** 125, **e** 81. **Quiz 4, a** 2, **b** 4, **c** 3, **d** 7, **e** 3. **Quiz 5, a** $\log_4 48$, **b** $\log_3 26$, **c** $\log_{10} 72$, **d** $\log_5 42$, **e** $\log_6 110$.

Percentages and proportions p 102: **Quiz 1, a** 3, **b** 308, **c** 392, **d** 118,404, **e** 86.4 billion. **Quiz 2, a** 32%, **b** 16%, **c** 43%, **d** 16%, **e** 6%. **Quiz 3, a** 0.2 & $\frac{1}{5}$, **b** 1 & 1, **c** 0.99 & $\frac{99}{100}$, **d** 0.05 & $\frac{1}{20}$,

e 0.04 & $\frac{1}{25}$. **Quiz 4, a** 5% increase, **b** 23% increase, **c** 32% increase, **d** 18% decrease, **e** 142% increase. **Quiz 5, a** $\frac{5}{6}$ kg flour & $\frac{1}{6}$ kg water, **b** Jog 2 miles & Walk 1 mile, **c** 50g eggs & 150g butter & 300g flour, **d** 40ml cordial & 320ml water, **e** 200g eggs & 80g onions & 160g potatoes. **Quiz 6, a** \$231.85, **b** \$90.05, **c** \$18.58, **d** \$2806.79.

Algebra page 110: **Quiz 1, a** $a = 2s$, **b** $c = \frac{2}{p}$, **c** $h = \frac{1}{4}w + \frac{1}{2}$, **d** $p = 4d + w$, **e** $F = \frac{9c}{5} + 32$. **Quiz 2, a** 8, **b** 2, **c** $\frac{1}{8}$, **d** 14, **e** 72. **Quiz 3, a** $4a$, **b** $3b + 1$, **c** $3x + 2y$, **d** $6x - 3a$, **e** $x + 5z + 2y + 2$. **Quiz 4, a** $4x + 4z$, **b** $2x + 8$, **c** $x^2 - x$, **d** $x^2 - 2xy$, **e** $2x^2 - 4xy$.

Equations page 117: **Quiz 1, a** 6, **b** 7, **c** 6, **d** 8, **e** 28. **Quiz 3, a** $x = 3$, **b** $x = 12$, **c** $x = 3$, **d** $x = 2$, **e** $x = 7$. **Quiz 4, a** $x = 2$, **b** $x = 5$, **c** $x = 8$, **d** $x = 0$, **e** $x = 3$. **Quiz 5, a** $x < 3$, **b** $x < 12$, **c** $x < 3$, **d** $x < 2$, **e** $x < 7$.

Angles page124: **Quiz 1, a** $\frac{1}{2}$, **b** 1, **c** $\frac{1}{4}$, **d** $\frac{3}{4}$, **e** $1\frac{1}{2}$. **Quiz 2, a** Acute, **b** Acute, **c** Acute, **d** Obtuse, **e** Reflex. **Quiz 3, a** 285°, **b** 210°, **c** 120°, **d** 37°, **e** 34°. **Quiz 4, a** 90°, 90° & 90°, **b** 135°, 45° & 135°, **c** 159°, 21° & 159°, **d** 58°, 122° & 58°, **e** 4°, 176° & 4°.

Triangles page 132: **Quiz 3, a** 60°, **b** 15°, **c** 40° at A & 70° at C, **d** 55° at A & 55° at C. **Quiz 4, a** 1cm^2, **b** 2cm^2, **c** 6cm^2, **d** 2.5cm^2, **e** 9cm^2.

Circles page140: **Quiz 1** Exact answers (to 1 decimal place), **a** 12.6cm, **b** 6.3cm, **c** 25.1cm, **d** 12.6cm, **e** 6.3cm. **Quiz 2, a** 15.71cm, **b** 31.42cm, **c** 3.50 miles, **d** 15.92km, **e** 0.70mm. **Quiz 3, a** 78.54cm^2, **b** 19.63cm^2, **c** 14.25mm^2, **d** 3.56mm^2, **e** 0.36mm^2. **Quiz 4** Each to 1 decimal place: **a** 1.3 miles, **b** 4.1mm, **c** 8.2km, **d** 10.2cm^2, **e** 7.5m. **Quiz 5** Each to 1 decimal place: **a** 28.3cm^2 & 50.3cm^2, **b** 22.0cm^2, **c** 25cm^2 & 19.6cm^2, **d** 5.4cm^2, **e** 77.1 cm^2.

Area and volume p 148: **Quiz 1, a** 28m^2, **b** 5000m^2, **c** 600cm^2, **d** 16cm^2, **e** 144. **Quiz 2, a** 125,000cm^3, **b** 12,000m^3, **c** 268.1cm^3 (to 1 d.p.), **d** 31,415.93cm^3 or 0.03m^3 (to 2 d.p.), **e** 3141.6cm^3 (to 1 d.p.). **Quiz 3, a** 15,000cm^2 (including the top face), **b** 3400m^2 (including the bottom face), **c** 201.1cm^2 (to 1.d.p.), **d** 6911.50cm^2 or 0.69m^2, **e** 993.5cm^2 (Curved surface only).

Polygons and solids p 156: **Quiz 3** (To nearest degree), **a** 120°, **b** 129°, **c** 135°, **d** 144°, **e** 179°. **Quiz 4,** Tetrahedron: 4 corners, 6 edges, 4 faces, Cube: 8 corners, 12 edges, 6 faces, Octahdron: 6 corners, 12 edges, 8 faces, **d**ocecahedron: 20 corners, 30 edges, 12 faces, Icosahedron: 12 corners, 30 edges, 20 faces.

Pythagoras' theorem p 166: **Quiz 1** (Each to 1 decimal place), **a** 5 miles, **b** 1.4cm, **c** 3.6m, **d** 7.8mm, **e** 17km. **Quiz 2** (Each to 1 decimal place), **a** 2.8mm, **b** 3.5 miles, **c** 9.1cm, **d** 8.5km, **e** 60 inches. **Quiz 3** (Each to 2 decimal places), **a** 5 metres, **b** 67.68cm, **c** 10.05km, **d** 147.41m, **e** 500m. **Quiz 4 b** 6cm^2, 49cm^2, 9cm^2, 16cm^2, 25cm^2, **c** 2.5cm, 6cm, 6.5cm, **d** 8cm, 15cm, 17cm **Quiz 5, a** 25, **b** 17, **c** 41, **d** 60, **e** 12.

Trigonometry p 174: **Quiz 1** (Each to 2 decimal places), **a** 0.31, **b** 0.48, **c** 0.98, **d** 0.52, **e** 0.88. **Quiz 2** (Each to 2 decimal places), **a** 3.53 miles, **b** 2.93cm, **c** 0.67 metres, **d** 8.49 nm, **e** 37.31mm. **Quiz 3, a** 3.5 miles, **b** 2.0cm, **c** 0.9km, **d** 88.8mm, **e** 4.7μm. **Quiz 4** (Each to the nearest degree), **a** 53°, **b** 10°, **c** 54°, **d** 59°, **e** 26°.

Coordinates p 182: **Quiz 1, a** (1, 3), **b** (−2, 3), **c** (−3, 2), **d** (−2, −3), **e** (4, −3). **Quiz 3, a** 5, **b** 13, **c** 2.83 (to 2.d.p.), **d** 3, **e** 12.81 (to 2.d.p.). **Quiz 5** (Each to 2.d.p.), **a** 1.73, **b** 1.41, **c** 3.16, **d** 4.12, **e** 5.48.

Graphs page 188: **Quiz 1, a** $y = x + 2$, **b** $y = 2x$, **c** $y = -x$, **d** $y = 4x$, **e** $y = 2x + 1$. **Quiz 3, a** $m = 3$ & $c = 0$, **b** $m = 3$ & $c = -1$, **c** $m = -1$ & $c = 0$, **d** $m = -1$ & $c = 1$, **e** $m = \frac{1}{3}$ & $c = 1$. **Quiz 4, a** $y = 5x + 4$, **b** $y = -x - 1$, **c** $y = \frac{1}{2}x + 2$, **d** $y = -3x + \frac{1}{2}$, **e** $y = 8$.

Statistics p 197: **Quiz 1, a** 41kg, **b** 1.0m, **c** 45.75, **d** 33db, **e** $1.4\dot{6}$. **Quiz 2, a** 42kg, **b** 0.9m, **c** 20, **d** 31db, **e** 1. **Quiz 3,** The mode is 1 & the median is 2. **Quiz 4,** The first quartile is 1, the third quartile is 3. The interquartile range is 2. The 99th percentile is 5.

Probability p 205: **Quiz 2, a** $\frac{1}{13}$, **b** $\frac{1}{2}$, **c** $\frac{5}{12}$, **d** $\frac{1}{4}$, **e** $\frac{2}{5}$. **Quiz 3, a** Yes, **b** Yes, **c** Yes, **d** No, **e** Yes. **Quiz 4, a** $\frac{1}{4}$, **b** $\frac{1}{26}$, **c** $\frac{1}{169}$, **e** $\frac{1}{26}$. **Quiz 5, a** Yes, $\frac{7}{26}$, **b** Yes, $\frac{2}{3}$, **c** No, **d** Yes, $\frac{1}{18}$, **e** Yes, $\frac{1}{1352}$.

Charts p 212: **Quiz 1, a** 90°, 180° & 90°, **b** 240°, 60° & 60°, **c** 72°, 216°, 36° & 36°, **d** 225°, 45°, 45° & 45°, **e** 252°, 72° & 36°. **Quiz 3** (Each to the nearest degree), **a** 164° & 196°, **b** 15°, 15°, & 330°, **c** 96°, 180°, 24° & 60°, **d** 248°, 34°, 72° & 6°, **e** 168°, 123°, 40° & 29°.

Index

PICTURE CREDITS

iStock: 4, 11, 66, 86, 140, 188, 205
Shutterstock: 18, 24, 35, 44, 73, 80, 95, 102, 110, 124, 174, 182, 197, 212
Thinkstock: 50, 57, 117, 148, 156, 166
Patrick Nugent: 132

Quercus Publishing Plc
21 Bloomsbury Square, London, WC1A 2NS

First published in 2011

A catalogue record of this book is available from the British Library

UK and associated territories: ISBN 978 0 85738 584 0
US and associated territories: ISBN 978 1 84866 165 3

Edited by Mairi Sutherland
Designed and illustrated by Patrick Nugent
Typesetting by Lapiz

Printed and bound in China

10 9 8 7 6 5 4 3 2 1